PATENTS

An Introduction for Engineers and Scientists

PATENTS

An Introduction for Engineers and Scientists

S. R. CAPSEY, C.P.A.

LONDON

NEWNES-BUTTERWORTHS

THE BUTTERWORTH GROUP

ENGLAND
Butterworth & Co (Publishers) Ltd
London: 88 Kingsway, WC2B 6AB

AUSTRALIA
Butterworths Pty Ltd
Sydney: 586 Pacific Highway, NSW 2067
Melbourne: 343 Little Collins Street, 3000
Brisbane: 240 Queen Street, 4000

CANADA
Butterworth & Co (Canada) Ltd
Toronto: 14 Curity Avenue, 374

NEW ZEALAND
Buterworths of New Zealand Ltd
Wellington: 26–28 Waring Taylor Street, 1

SOUTH AFRICA
Butterworth & Co (South Africa) (Pty) Ltd
Durban: 152–154 Gale Street

First published in 1973 by
Newnes-Butterworths, an imprint
of the Butterworth Group

© Butterworth & Co (Publishers) Ltd, 1973

ISBN 0 408 00104 6

Printed in England by
Willmer Brothers Limited, Birkenhead
Bound by C. Tinling & Co, Liverpool

Preface

This book is an expanded version of a booklet produced for use within the company for which I work, and which has attracted favourable comment. The purpose was to explain to engineers why patents exist, and consequently why patent agents and patent departments exist. It has now become apparent that a wider market exists for such an explanation.

I would like to acknowledge the assistance I have received from numerous earlier published articles and a few books, and also from my colleagues. In this respect I am especially indebted to Dr Roger Bones for helpful criticisms of the original document, and to Jim Halliwell for suggesting that publication on a wider scale was worthwhile.

For much of the information in Chapter 1 I am indebted to a paper by M. Frumkin, *The Early History of Patents for Invention*, which was read on 12 November 1947 to a joint meeting of the Chartered Institute of Patent Agents and the Newcomen Society, and was printed in the Transaction of the Chartered Institute of Patent Agents, Vol. LXVI, 19–69, 1947–48.

Finally, I would like to thank Standard Telephones and Cables Ltd. for giving me permission to publish the book.

S.R.C.

Contents

1

Historical Survey

The general belief has been that invention patents originated in and were copied from England, as indicated by Goethe's comment that the invention patent 'originated among the English, who draw profit and advantage out of everything'. However, other countries had the idea of monopolies of invention at a very early date.

In antiquity the general atmosphere was pessimistic in that Greek legend talked of the deterioration from Golden Age to Silver Age to Bronze Age and so on. This was accompanied by the general feeling that practical arts and inventions were undignified; thus Plutarch praised Archimedes for not describing his practical inventions. Exceptions were of relatively little importance, one example beng the so-called 'food patents' of Sybaris, where a cook or confectioner who produced a new dish had a six months' monopoly of it. The idea that progress was towards, and not away from, a Golden Age was a product of the Renaissance, and as will be seen, patents are legalised progress.

Monopolies as such existed as long ago as the Byzantine Empire but they were not for inventions. One example of the early use of privileges for technology occurred in 1331 when Edward III wished to attract Flemish weavers to England. He issued letters of protection, analogous to a passport but not giving exclusive rights. The bait offered was a promise of 'good beer, good beds and still better company, English girls being renowned for their beauty'. This worked so well that the Flemish authorities had to make agreements to prevent the excessive 'export' of weavers.

Then in 1449, Henry II granted to John Utynam, who came to England to make stained glass windows for Eton College chapel, a monopoly of stained glass manufacture. This had been long practised on the Continent, but not in England, so here we have a monopoly

1

granted in return for introducing a new technology into this country. Similar monopolies for introducing new ideas into the realm were also granted in other countries.

Some of the Italian states had early grants which to some extent resembled invention patents, the term invention patent meaning a grant of a monopoly in an invention for a limited period to the inventor or the owner of the invention. Then in Florence there were examples of early privileges relating to the encouragement of industries, and one of these was probably the earliest which could be called an invention patent as now understood. This, in 1421, was granted to Brunelleschi for a barge with hoisting gear for transporting marble used in building. However, in spite of such grants of privilege, no patents system was set up. Venice also granted what we would now call patents, and in 1474 this lead to the Venetian senate voting the first patent law by a heavy majority. The main portion of this is:

'It is enacted by the authority of the present council that whoever will make in the City any new and ingenious artifice, not made previously in our state, will be obliged to register it at the office of our provediters of the Commune, as soon as it will be reduced to perfection so that it will be possible to use and apply it. It shall be forbidden to anyone else in our land and place to make any other artifice to the image and similarity of that one without consent and licence of the author during the term of ten years. And if nevertheless someone did it, the aforesaid author and inventor would be free to cite before every office of this City, and the said who would have imitated would be compelled to pay one hundred ducats, while the artifice would be immediately destroyed. But our Government will be free, at its complete discretion, to take over and use for its needs any of the said artifices and instruments, under the condition, however, that others than the authors may not employ them.'

This was not strictly applied, and some of the patents lasted more than 10 years—usually in multiples of five. Note in this Venetian law the preserving of the State's rights to use the invention.

Knowledge of the Venetian patent system, plus a flourishing patent system at Antwerp, lead to suggestions that such a system be set up in England, and patents for inventions began to be granted in the sixteenth century. At this time most monopolies were of a more unsatisfactory kind, being trade monopolies, usually granted to a sovereign's favourites as rewards or bribes, but more of this later. Many of these monopolies were for 10 or 20 years, rather than the traditional English term of 14 years (two generations of apprentices) which came later.

Many of the early English patents were to foreigners, e.g. to Francis Berty (1565), an Italian who patented in other countries, including Scotland where a patent system got under way early. This was soundly based: thus a 1606 patent for a coal mining invention included a clause that the content be 'ane work and ingyne nocht knowin in this Kingdome at na tyme before'. Early Scottish patents included one in 1694 for 'ane engine for writing, whereby five copies may be done at the same time', and in 1699 'wooden kettles and cauldrons which could abide the strongest fire'. The term was 19 years, as in many farm leases.

In Ireland, the first patents, as in many countries, were for glass-making, e.g. one to Woodhouse in 1586. The art of glassmaking seems to have spread abroad from Venice.

Going back a little in time we will look at the Antwerp system mentioned above. Here there was a well established patent system in the sixteenth century and as early as 1551 we find a patentee criticised for not working his patent. This is one historical precedent for the inclusion in many countries' patent laws of provisions making it obligatory for a patentee to work his invention or be willing to grant licences to others to work it. There were also patents valid in Brabant and Liege, and a flourishing Dutch patent system before 1550.

France also had an early patent system, and here it was usual practice to examine the inventions for which a monopoly was sought and also to perform experiments. Thus in 1609, Le Courè and Thouyn petitioned for patents for furnaces which they claimed they had invented. A practical examination before the minister Sully was imposed to decide the issue. In spite of all these rather advanced ideas, the French system was apparently without any formal legal basis.

Many other countries were early in the field of patents for inventions, these including German States. In 1535 the Margrave of Brandenburg-Onalzbach, who owned Tarnowice in Silesia, granted an eight year monopoly for a pump to be used in mines. Austria, Poland, Berne, Zurich and Russia also granted patents in the sixteenth and seventeenth centuries. One interesting point about the early Dutch system was that in the sixteenth and early seventeenth centuries written descriptions of the inventions were consistently required. This was not always so in other countries.

Having looked briefly at the early history of patents in other countries, we can now look in a little more detail at England. The term 'letters patent' literally means 'open letters', and referred to grants of privilege issued under the Great Seal. These included grants of dignities such as letters patent of nobility, appointments to certain offices of state and monopoly rights.

As already mentioned, monopoly rights were much abused in this country and had led to complaints in Parliament as long ago as 1347, when a foreigner named Tidman had an exclusive right to export Cornish tin without paying tax. This abuse apparently got worse, so that in her last Parliament, Queen Elizabeth I was forced to revoke a number of her letters patents which related to such offensive monopolies. Such patents, however, were granted in the early years of James I's reign, including one giving the sole right to make cigars. In view of the King's views on smoking, this monopoly is rather surprising.

The agitation caused by such abuse of monopolies led Parliament in 1623 to pass the Statute of Monopolies, which made all monopolies illegal except for those relating to 'Manners of New Manufacture'. The word 'new' in this exception is significant, and this part of the Statute of Monopolies was the foundation of our patent system. In fact the words of Section 6 of this Statute still appear in the definition of invention in the Patents Act 1949. This definition reads as follows:

' "invention" means any manner of new manufacture the subject of letters patent and grant of privilege within section six of the Statute of Monopolies and any new method or process of testing applicable to the improvement or control of manufacture, and includes an alleged invention'.

In the above quotation from Section 101(1) of the Patents Act 1949, the reference to methods of testing was newly inserted in 1949.

During the years which followed the enactment of the Statute of Monopolies, patents were granted fairly frequently and there were actions for infringement of these patents. These included a famous one in which James Watt successfully sued Cornish mine owners for infringing his patent on a condensing steam engine (*Boulton and*

Watt v. Bull, 1795). However, it was very expensive to get patents, and separate patents were needed for England (which included Wales), Scotland and Ireland. With the approach of the Great Exhibition in 1851, Acts were passed to produce a United Kingdom patent system, and also to reduce the costs to the applicant.

The patent specification reached something like its present form with the provision in the 1883 Act that the specification which described the invention must end with claims defining the invention. The significance of these claims will be discussed later. In 1905, after an investigation which showed that a large proportion of the granted patents were invalid in view of what was known before they were applied for, the Patent Office was instructed to carry out searches in the prior art to see if anything the same as what was claimed as the invention could be found. This put the patent system on the basis on which it still stands today.

In the intervening years, a number of amendments and revisions have been made to the patents legislation, leading to the Patents Act 1949, on which the patent system is now based. Thus it is now more than 20 years since any substantial alterations were made. A report has recently been issued by the Banks Committee, which was appointed in 1967 to investigate changes needed to the patent system. Some of their recommendations will almost certainly be implemented, so this report will be discussed later. Other circumstances which will alter our patent system are Britain's entry into the Common Market, and various international arrangements for patent co-operation.

2

What is a Patent?

A letters patent for an invention arises as a result of a contract between the inventor and the Crown, whereby the Crown grants to the inventor a monopoly for his invention for a limited period in return for the inventor disclosing his invention. In practice the contract is usually between the Crown and the inventor's employer, and annual fees have to be paid by the patentee to keep the patent in force. The merit of this disclosure of the invention is that the risk of the same invention being produced independently and successively by several people is minimised. However, nothing can prevent several people having the same idea at about the same time. This does happen, especially in fields in which development is rapid.

At this point it should be emphasised that a patent is not granted for an idea or principle as such, but for some product, or machine, or process using that idea. Scientific principles and laws of nature, of course, are specifically not patentable. This follows from the emphasis in the Patents Act on 'manners of new manufacture'. This, like many legal terms, is hallowed with age, derived as it is from the Statute of Monopolies (1623). Crudely speaking, a 'manner of manufacture' is something you can sell (called by Mr. Justice Morton in a famous case, G.E.C.'s application, 60 R.P.C. 1*, 'vendible product'), or a machine or process for making it. Some methods of testing are also patentable if applicable to the improvement or control of manufacture. Naturally a new machine for testing is patentable.

Exclusions from what is patentable include such ideas as novel filing or accounting systems, camouflage systems, musical notation, ideas which are mere arrangements of old and well known objects,

* *Reports of Patent, Design, Trade Mark and Other Cases,* Volume 60, page 1. See footnote on page 28.

many medical treatment ideas, and horticultural and agricultural ideas. On the subject of what is patentable there is an extensive background of judicial decisions, but the concept of what is patentable is changing, usually to include types of inventions which formerly were not regarded as being patentable.

An invention, or to be more accurate legally, a patent for an invention, is a form of industrial property, and industrial property is similar in many ways to private property. Thus the owner can sell all or a part of his private property, and equally the owner of industrial property can sell all or a part of it. Again, just as an owner can rent or lease private property for a period, so the owner of industrial property can grant licences to others to use it. Normally the consideration for such a licence is a royalty. Yet another point of resemblance is that the owner of private property can sue trespassers, while with industrial property the owner can sue for infringement of his rights. Both forms of property are part of the owner's estate and would on his death pass to his next of kin or the person specified in his will. In the case of industrial property owned by a company, this means little as the company outlives most people.

In the case of private property the document which specifies the ownership defines the boundaries of that property. Similarly in the case of the patent specification for an invention, we have the claims which have already been mentioned. These are brief statements which define the boundaries of the monopoly of the patent. In both cases the documents are subject to the laws of interpretation of documents. Thus it will be seen that a patent specification is both a technical document and a legal document. This is an important consideration and is the reason for the existence of patent agents.

A British patent, once granted, lasts for 16 years from the date on which the Complete Specification (see Chapter 7) was filed at the Patent Office, provided that the owner pays the annual renewal fees which are due from the fifth year on.

3

The Ownership of the Invention and the Patent Therefor

The majority of patents granted nowadays are for inventions made by employees of companies, relating to ideas relevant to the employer's business. The basic law in this country as applied to an employee who makes an invention is that if the invention concerns the employer's business, then it belongs to the employer. This normally applies even though the invention was dreamed up either wholly or in part in the employee's own time. The principles behind this apparently harsh approach are two-fold. Firstly an employee usually acquires knowledge in the course of his work which either leads him to the invention or helps him to develop it. Secondly any other ownership of such an invention would be inconsistent with the obligations of good faith which should exist between an employee and his employer.

The second point follows a judicial statement in a leading case, in which a judge said:

'It is an implied term in a contract of service of any workman, that what he produces by the strength of his arm or the skill of his hand or the exercise of his inventive faculty shall become the property of his employer.

'If the employment is of a designer, that which he designs is thus the property of the employer, which he alone can dispose of. If it is patentable, it is for the employer to say whether it shall be patented and he can require the employee to do what is necessary to that end. And if it is patented in their joint names, the employee holds his interest as trustee.'

The above arises from considering work and inventions from

engineers and others who, by virtue of their job, can reasonably be said to be 'paid to invent', i.e. it is part of their job to have ideas relating to the improvement of the company's business. In the case of a 'shop floor' employee who has an idea relating to the company's business which might be patentable, the rights in it would normally belong to the company, but in many companies the employee would get some specific reward for that idea, for instance from the company's suggestions scheme.

The employer–employee relation discussed above assumes that there is nothing in the employee's contract of service to alter that relation. Thus some companies pay cash bonuses to inventors, in one case a small sum of money on filing the application plus additional sums if patents are filed overseas, these bonuses being additional to the salaries. Although patents are usually filed in the names of the employers, it is possible (as sometimes occurs) to file jointly in the names of the company and the inventor(s). The latter can cause administrative difficulties since for some legal purposes every applicant for a patent has to sign documents and approve of what is done with the patent. This is inconvenient where an inventor has died, or left the company and cannot be contacted.

The above relation is reasonably satisfactory, and is in fact reasonably fair, with companies of the traditional type whose range of activities is fairly narrow. However, in these days of multi-product groups of companies, the enormous range of activities of the group could lead to difficulties where an employee has an invention well outside the scope of his work but within the range of the group.

It will be reasonably clear that when an employee has an invention outside his work he would be wise, in all cases, to check with the company as to whether he can legally exploit it. Most companies treat such matters reasonably. Although there has been a vast amount of litigation between inventors and employers, this has often been due to carelessness, in that the rights of the parties have not been properly defined, or to obstinacy on one or both sides.

This Chapter is a convenient point to refer to patents which are granted for inventions 'communicated from abroad'. Such patents reflect the desire in the old days to bring new technologies into the realm, such as in the case of the Flemish weavers mentioned in Chapter 1. In such a case a person resident in this country files a patent application in his own name, and usually states in the application documents that the invention was communicated from a named person or company abroad. This was originally important due to difficulties of travel, but in modern times it is usually used as a convenience when, for instance, there is difficulty in getting signatures. Thus a patent agent will often file an application in his own name, as a communication from an overseas client, and subsequently execute a formal assignment whereby the rights in the patent are assigned to the client.

4

The Value of Patents

Unfortunately, many engineers do not approve of patents, which in their eyes are merely an excuse to force them into work which is separate from the real purpose of their lives—engineering. Hence it is necessary to persuade them that patents have a value, and that a company can be greatly inconvenienced without them. It is also necessary, in many cases, to bring home to engineers the central fact that you cannot get a patent until someone has invented something.

It has been shown already that a valid patent with strong claims to define its invention enables the patentee to prevent someone else from using the invention, or to force him to pay for the privilege of using it. Thus one obvious use for a patent is to prevent competitors from copying what the patent covers. Therefore, when a company introduces a new product, it is desirable to examine it to see if it contains any patentable ideas. This calls for discussion with the patent agents. Even a relatively narrow patent on a product which is being marketed or is going to be marketed may justify its cost if it prevents copying by a competitor. It may cause the competitor to design a viable alternative which forces him to spend money on development, and also, if it produces a saleable alternative increases the sum total of technical knowledge. Thus, we have the 'private' benefit in that the competitor has been driven to design around the patent, and the 'public' benefit in that he has introduced an alternative to the market.

At this point the matter of infringement of one's own patents should be mentioned. Some companies are in the fortunate position of employing people whose duties include specifically checking competitors' products in search of infringements. It is also possible to use outside help. However, in most cases detection of infringement is to some extent a random matter. When this is so, those best fitted to detect infringement are probably engineers and salesmen—assuming that they know something about their company's patent holdings. This would apply especially when a competitor is competing successfully—possibly even underselling—with something the same as or similar to what the company sells. In such a case it is useful to investigate whether the competitor is infringing a patent, in which case the company's patent agent (whether an employee or an outside consultant) should be notified. He will need evidence, usually information about the competing product which is believed to infringe.

A related circumstance which is of interest is where a competitor is selling something similar to what one's own patent covers, but which just avoids infringement. This could well mean that the competitor had been forced to design around the patents, which would often justify their existence. Hence, retaining them by paying the renewal fees would be worthwhile as it would restrict the competitor's freedom of action.

Patents can also be exploited by the grant of licences, which can be a lucrative practice. Thus one could grant a licence to another company who wish to use one's inventions or who are detected infringing one's patents under one or more patents, in return for payment of royalties. The licensee might pay 3 per cent on the selling price of the article produced. Alternatively, reciprocal arrangements can be entered into, whereby the two sides grant each other licences. Any actual cash payment would depend on how many patents and of what quality the two sides owned.

Related to patent licensing agreements there are 'know-how' agreements which relate to manufacturing techniques which are not patentable, or which the owners do not wish to publish. These agreements are often strengthened if one or more patents can be included in the agreement. Other agreements in which patents are involved may be basically commercial agreements, which may be more easy to enforce when patents are used to back them up.

An important aspect of patent licensing arises from the provision in the Patents Act whereby a patentee can have his patent endorsed 'licences of right'. A patent can be so endorsed at any time during its life, and the patentee is indicating thereby a willingness to grant a licence on reasonable terms to any applicant. In return, the annual renewal fees are halved. The process of obtaining the endorsement is relatively simple: the appropriate form (Patents Form 42) is filed at the Patent Office, duly stamped with the prescribed fee, and supported by a declaration by an officer of the company that there is no agreement in existence which would be contravened by such an endorsement. Thus the endorsement has adequate safeguards for any existing licensees. Unlike the situation in some countries, an endorsement 'licences of right' can be cancelled by the patentee, in which case it is necessary to pay the amounts saved by the halved renewal fees. Such cancellation is sometimes done when a licence is granted if the licensee wishes to be the sole licensee.

Both the application to endorse a patent 'licences of right' and a later one (if any) to cancel such an endorsement are advertised by the Patent Office in the Official Journal (Patents), and can be opposed by any interested parties. This helps to provide safeguards for any existing rights. In some cases the facility to endorse 'licences of right' is a useful intermediate stage between keeping a patent in force at the full fee and allowing it to lapse.

It is necessary here to indicate the distinction between a sole licensee and an exclusive licensee. If you grant an exclusive licence, you are excluding yourself from the use of the patented invention, whereas a sole licence makes the licensee the only person other than the patentee who can use the invention. Another interesting distinction is that the exclusive licensee can, by virtue of the provisions of the Patents Act, sue for infringement if the patentee is unwilling to do so.

It should be noted that one can grant two or more exclusive licences, e.g. basing them on technical or geographical distinctions. Thus, if the invention relates to electronic circuits, it may be possible to grant to one company an exclusive licence to use the invention in its transistorised version, and to grant someone else an exclusive licence to use the vacuum tube version. In the case of geographical limitations, licences can be granted to use the invention in different areas. Similar considerations, of course, apply in the case of sole licences, and indeed in licences generally.

A use of patents which is allied to the reciprocal licences mentioned above is to enable the owner to join a patent pool. This gives an income to the members of that pool which is based on the numbers and/or strengths of the patents in the pool, and naturally gives all members rights under all of the patents in the pool. A typical arrangement in a patent pool is that royalties from licences granted on the patents in the pool, either all of them or selected ones, are put into a common fund, and the members of the pool each take a share of these royalties based on their holding in the pool. In addition, the members pay royalties in respect of their own manufacture in the field. Then by owning, say, 10 per cent of a batch of patents one gets the right to use all of them, and also to receive some royalties. Such pools can be valuable where a new field of technology is developing, and the important patents are spread among several companies.

A factor which must be mentioned is the alleged suppression of inventions by patents sometimes referred to in certain organs of the Press. What is apt to be ignored is that the British Patents Act includes provisions which forbid the misuse of patents. If the patentee is not using the invention, or is failing to meet the market at reasonable cost, or is relying entirely on imports, or is failing to supply a useful export market, an interested party can apply to the courts for the grant of a compulsory licence. This is not a free licence and, in general, the provision is not much used, because most people would rather make money from royalties in the above circumstances than not. Again, one should note that once a patent has been published the contents of it can be read by anybody, so that when it lapses (either of old age or through renewal fees not being paid) its contents are in the public domain.

A special case is that of patents relating to food production, medicines and certain surgical and related machinery. In such cases compulsory licences are obtainable as of right, as long as the applicant satisfies the court that he can adequately work the invention. However, this special case will disappear if one of the recommendations of the Banks Committee is implemented.

Patents are sometimes useful in helping to sell the patentee's products, as some customers, especially Government departments, are more willing to purchase goods if the seller has patents or patent applications.

An important aspect of patenting is prestige, where the patents relate to important inventions. One of these was pulse code modulation, invented before World War II by the late A. H. Reeves, which has been celebrated by a British postage stamp.

The life of a patent is 16 years from the date of filing the Complete Specification (see Chapter 7), provided that the renewal fees are paid. However, if the invention is a valuable one for which the owner has not been adequately remunerated, the patentee can get an extension of the term of a patent on application to the High

Court. An extension is also possible where the patentee has been prevented from working the patent by reason of a state of war involving this country. In this case the application can be to the High Court or the Patent Office, the latter being preferable as it is cheaper.

Extensions are free of charge when granted and it is possible to get both sorts on the same patent. An example of this is the patent on a fundamental feature of colour television as used in this country. This is owned by EMI, the inventor being Georges Valensi, and the patent was obtained before World War II. By the time it expired it had lasted over 30 years and was the subject of patent infringement action. In this action, Mr. Justice Graham held that the patent was valid and infringed, and the defendants have appealed, decision in the lower Court having been reversed by the Court of Appeal.

Another valuable invention which led to litigation is that of printed circuits; the main inventor was Paul Eisler and the British patents from his work were owned by Technograph. This company sued a number of companies under these (now expired) patents, and in the one case, against Mills and Rockley, which went to the Courts, they won in the House of Lords.

To re-iterate, the patent system cannot be used to suppress inventions, since the acceptance of a patent application leads to its publication in printed form. Hence, as soon as the patent lapses, either at the end of its term or earlier if the owner ceases to pay the renewal fee, the information it contains is public property. One example which used to be quoted in this context is the so-called universal match. Two patents on this were granted in this country, No. 369621 (F. Ringer) and No. 374602 (R. König) in the 1930s. These were duly published and have long since lapsed, both expiring in 1936, when the patentees did not pay the renewal fees. The inventions they relate to are now public property and can be used by anybody interested. The author recently saw an advertisement in a Sunday newspaper for a 'universal match', so someone could be using one of these inventions.

Another bar to suppression of this sort is that the Patents Act contains provisions against abuse of patent rights. These are rather technical, but what they amount to is that if the patentee is not using the invention, or not adequately supplying the market, or supplying the market wholly by import, or not supplying an export market, and someone who seeks a licence is refused one, then that person can apply for a compulsory licence. In the medical field, all patents can be compulsorily licensed, with certain safeguards, after three years from grant.

These compulsory licence provisions are used to some extent in the medical and related fields, but outside them they are little used. This is probably because they exercise a salutory restraint on the less respectable patent owners.

5

Defensive Patenting

This is to some extent an aspect of the subject of the previous chapter. It will not have escaped the notice of anyone involved with patents, that many granted patents are of doubtful validity, usually because their contents so closely resemble what has gone before that they would be held by a court to be invalid on the ground of obviousness. However, it is often sound commercial sense to get such patents. Even if they are of dubious validity, it may be expensive to get them declared invalid since this calls for legal action. In many cases, of course, such a use of a bad patent is in truth a misuse of the patent system.

If a company owns an idea which is commercially valuable and does not patent it, there is the risk that someone else will have the same idea and get a patent which could be a major nuisance if valid, and quite an embarrassment if not! Many large companies play for safety and get a patent on such an idea, even if it is known that it would be of dubious validity at best. Such weak patents do not, however, always arise from this motive only, and there are other reasons for getting patents in this 'limbo of possible or probable invalidity', such as:

(1) the technology may be commercially valuable, and/or may be linked to other patents on an important development;

(2) the resulting patent may have nuisance value to a competitor and thus be a useful bargaining tool;

(3) it was alleged in the past to be nearly as cheap to publish by getting a British patent as to publish in a journal, although with certain specialised journals which have appeared recently this consideration is of less significance;

(4) publication by a patent can be delayed by as much as three and a half years from the application date, which gives one

time to reflect on its importance, and the application date fixes when you thought of it;

(5) at its date of conception the commercial value may be difficult to assess, so one tends to file in the interests of safety;

(6) the patent names the inventors who gain some public recognition on publication;

(7) where a market is expanding rapidly, it may be desirable to file patents on everything, whatever the inventiveness. Thus a senior executive of Rank-Xerox was heard to say that they 'patent everything in sight'.

When one has a new idea which looks as if it would at best produce a weak patent, there are three alternatives:

(a) publish immediately and thus inform a possible competitor about it, with no way to stop him from using it,

(b) patent the idea,

(c) do nothing, in which case there is the risk of a competitor getting the same idea independently, or learning of it as a result of your own staff leaving.

The general tendency with new ideas which are of potential commercial value is to apply for a patent.

On the general point of the value of patents, a press statement by AMP of Great Britain Ltd. was quoted in an issue of *Electronic Engineering*. AMP referred to several examples where they had successfully quoted their patents against competitors and came to the conclusion that having patents is sound commercial sense.

6

Publication

As mentioned in an earlier chapter, to support a patent an invention must be new. This means that it must not be known to the public at the date on which the application was filed (or in the case of an application of overseas origin filed under an international agreement, on the priority date see Chapter 8). Any publication of the invention before the application's priority date, if proved, would render the patent invalid. This consideration is often difficult in a field of technology which is developing rapidly, as in such cases many ideas appear simultaneously or nearly so in several places. Again, an idea is sometimes 'in the air', so to speak, or crops up in conversation before any proper reduction to practice of the idea occurs.

Where the patent application is first filed in Great Britain, the important date is the British filing date. However, here are a few exceptions to the rule that prior publication invalidates a patent. The most important is that if an invention is disclosed at an exhibition approved by the Board of Trade occurring not more than six months before filing date of the application, then that disclosure will not invalidate the application for the patent, or a patent granted thereon. This assumes that the exhibition disclosure is by the inventor or one of his colleagues. A second exception is where the inventor reads a paper before a learned society in this country not more than six months before the patent application is filed.

It would be dangerous to rely upon the above two exceptions if one wished to file any overseas applications. In either case the disclosure could lead to some further publication which could be harmful. An exhibition catalogue or a 'hand-out' could contain enough information to render a later patent application in other countries invalid from prior publication. Furthermore, the catalogue or hand-out could, if it contained enough information, form an adequate prior publication to render a British patent application useless.

A similar risk exists if one reads a paper to a learned society, since copies of such papers may well circulate outside this country, and since reports of the meeting might circulate widely. In either case, publications caused by the paper could render a foreign patent application useless.

In the above two cases, if the invention is likely to be patented one should ensure (if at all possible) that a patent application is filed before the exhibition or the reading of the paper. For this reason, most companies which file patents overseas have strict rules about consideration of publicity material, exhibition material and employee's papers for patentable material. To many engineers and to the publicity department this seems a nuisance but it is essential to protect the company's interests.

Another form of disclosure which would not upset a later patent is communication of the invention to a Government department, or to someone authorised by a Government department to investigate it. This is really a special form of confidential disclosure and, of course, confidential disclosure to another person would not invalidate a patent application filed later. However, the confidential nature of the disclosure must be clearly accepted by both sides: an offer of sale to a customer, without the confidentiality being acknowledged by both sides, would normally not be regarded as confidential, and could thus kill any later patent application.

Disclosure of an invention in fraud of the inventor's rights, or in breach of an undertaking to maintain its confidentiality would also not upset a later British patent application. However, here also one must remember that such accidents can lead to publications which would spoil the chances of getting valid foreign patents.

Merely marking a document 'confidential' does not necessarily prevent a document from being regarded legally as a publication. In a recently decided case (1957), a document which described a process had been circulated to all members of a trade association, but was marked confidential to members of that association. The Court took the view that as the members formed a substantial part of the 'interested public', then the existence of the document constituted publication (J. R. Dalrymple's Application for a Patent, 1957, R.P.C. 499*).

One more example of disclosure which does not upset a later application is where the invention is worked publicly for 'reasonable trial and experiment' for not more than one year before the filing date. Examples of this are rare, as most people file something before they do any such public working. The only example the author can remember was in the USA many years ago. A man devised a new method of paving streets, and to test it he paved a street in his home town and at his own expense. When he was satisfied that his method was sound, he filed a patent application and eventually got a patent. An attempt to invalidate the patent on the basis of the use of the invention failed. It was held that its use was essential to establish the practicability of the invention, and was therefore reasonable trial and experiment.

In view of the limited exceptions mentioned above, it is always preferable to file a patent application before any publication, or demonstration, or public use of the invention. As already indicated, this is the reason for careful vetting by an employer of publicity material, etc., before its use is authorised. In a company with a patent department it is therefore usual to submit all publicity material, whether exhibition publicity or advertisement, and all technical articles, papers or books to the patent department. If such material includes anything patentable which has not been the subject of a patent application, approval for publication should be withheld until a definite decision as to whether or not to patent has been taken. If the decision is to patent, then delay clearance for publication until after the application is safely on file. This may well be a nuisance, especially where some last minute idea is to be used at a trade show, but it is essential to protect the company's commercial interests.

An important item to consider in this context is the problem of visitors to a company's works or laboratory. If the visitors are outsiders, as happens when a company has an 'open day', it is advisable to make sure (in collaboration with the patent department, or a patent agent) that one does not show anything which is not patented or contained in a patent application (unless it is not going to be patented).

* *Reports of Patent, Design, Trade Mark and Other Cases,* Patent Office, 1957.

The only way to safeguard a patent application is to make sure that it is filed before any publication or prior use of any sort takes place. This preferably includes filing before disclosure to a Government department, because the more people who know about an invention, the more there is a risk of accidental disclosure.

On the subject of Government departments, there is another reason for filing before discussion. If one uses an invention in a job wholly or partly Government funded, the Government has greater rights in the invention if it was developed after the contract date than before. Argument as to when the development occurred can usually be short-circuited if one has a fairly early filing date to refer to.

7

The Patent Specification

This chapter is based on the British patent specification but much of it applies to patent specifications of other countries. However, one point which applies in this country to a greater extent than elsewhere is the Court's attitude to a patent specification. They tend to regard it as a legal document, and to apply to it the very strict rules of interpretation of legal documents which have been built up over the years. By contrast, in some countries the Courts tend to interpret documents to some extent at least in the light of what they were intended to mean. Our practice has the merit of certainty but does tend to put too much emphasis on the actual words used.

In this country, and in some Commonwealth and ex-Commonwealth territories, one can file a Provisional Specification or a Complete Specification with the original application. The Act states that every specification, whether complete or provisional, 'shall describe the invention', while the Complete Specification 'shall particularly describe the invention and the method by which it is to be performed; shall disclose the best method of performing the invention which is known to the applicant and for which he is entitled to claim protection; and shall end with a claim or claims defining the scope of the invention claimed'.

In practice the main differences which these quoted passages from the Patents Act indicate are that the Complete Specification usually describes the invention more fully than the Provisional, and that the Complete Specification includes the claims which will be described below. If one initially files a Provisional Specification, a Complete Specification should be filed within a year; this can be extended by up to three months on paying a fine, but the application will otherwise lapse and will not be published.

A typical British Complete Specification starts with a general

C

introductory portion, including a first paragraph describing roughly the field to which the invention relates. This might be a simple one-sentence paragraph, such as:

'The present invention relates to automatic telephone exchanges, and especially to such exchanges in which communication connections are set up via crossbar switches.'

Such a paragraph is sometimes followed by a statement of current practice or prior art, with some of the disadvantages thereof, and a statement that the object of the invention is to overcome such disadvantages. Then would follow a statement of the invention, which usually (but by no means always) uses exactly the same wording as one or more of the claims. The reason for this repetition is given below. The introductory paragraphs can contain any technical definitions which the writer considers desirable, and may also include general descriptions of one or more embodiments of the invention, i.e. actual examples of how the invention is used, but these really should be in a preamble to the specific description.

The specific description is a description, usually with reference to one or more drawings, of a least one practical form, or embodiment to use the legal term, of the invention. The example or method described is by example only, and it would not usually be possible to avoid infringing the patent by working the invention in a different way. Thus, if the invention relates to electronic circuits, and the circuit actually described used vacuum tubes one would not necessarily avoid infringement by using transistors in an otherwise similar circuit. If one did, it could be because there was some feature of the circuit described whereby transistors could not be used, or because the claims had been badly written, or because the patent agent had been inadequately briefed by the inventor. This is a very important point: when one describes one's invention to the patent agent, one should remember that his knowledge is less. He will therefore often ask questions which seem very elementary, but for the sake of a good patent it is essential to answer them.

Since, as pointed out above, the description with reference to the drawings merely describes one or more examples of how the invention is put into practice it can be, and usually is, relatively specific. The detail given may be quite considerable, with alternatives mentioned. The advantages conferred by the use of the invention are also often given.

The description should be 'clear to one skilled in the art', and in general in this country that means a reasonably intelligent development engineer. It should also be a good technical write-up, not cluttered with legal jargon, a criterion which is not always met.

One type of patent specification which can cause difficulties is that for a 'chemical' invention. This poses problems which differ from those of mechanical inventions, and can be especially difficult when a new chemical compound is involved, as there may be difficulties in defining it accurately when (as can easily occur) the chemical formula is not known for certain. Here it is even more important than usual to keep the patent agent fully informed. In chemical-type inventions, it is usually essential if a relatively wide claim is needed to have several examples of putting the invention into practice. If the invention is a new process, it is desirable to describe several examples of that process. If there is only one example a Court may hold that the invention is legally limited to that one example. Hence, the patent agent will be asking such questions as: how can these percentages of ingredients be altered? What other ingredients could be used? What ranges of temperatures? It may even be necessary to do a few more experiments to answer these questions, but tiresome as they seem, if the invention is commercially valuable they are well worth the trouble.

The claims which have already been mentioned several times are the most important part of the specification from the legal point of view, so they merit a chapter of their own. However, it is worth stating that the main claim should be the best possible brief statement the writer can devise of what the invention is, and this is why the main claim is often repeated verbatim in the statement of invention. Obviously, if it is in fact the best possible such statement, then an alternatively worded claim might well be inferior. Further, if it differs to any extent this could lead the Court to say that there was ambiguity.

8

The Claims

In the previous chapters, claims have frequently been mentioned. Claims are clauses which appear at the end of the specifications, and which define, or should define, precisely the area of the monopoly claimed. They are not necessarily summaries of how the machine or process actually described works, although if they can do this without sacrificing the precision of definition just mentioned, no harm is done. From the legal point of view the claims are the most important part of the specification.

The broadest claim is usually the first one, and this sets out what the inventor or his patent agent believes to be the real invention. This should set out a combination of features, or occasionally a single feature, which is new, and which is not obvious when compared with the prior art. The point of this is that often a device is produced which is different from what has gone before, but which is so closely similar that it would have been obvious to one skilled in the art. The question as to whether some new device is or is not obvious is a matter of opinion, and when this question comes to be argued before the Court it is usually an expensive and complex matter.

Following from the need for a claim to be precise, it should only have a single meaning, which should be the meaning that the author intended it to have.

The claim should relate to what is usually called an 'operative combination', i.e. to some combination which will work. If it is so written that it includes a device which does not work then the claim is almost certainly invalid. This is one of the points which can make it difficult to write, and also makes it important to brief the patent agent fully. Again, when the inventor is asked to check the specification and claims in draft form, this point is one which he could usefully consider. Usually he will not be able to assess the claim's soundness as a legal statement, but he should be able to assess its technical accuracy.

In addition to the main claim, there are often one or more other claims of an independent nature, i.e. claims each of which stands on its own and does not refer back to an earlier claim. Two or more independent claims may be essential where there are different sorts of devices using the invention which can be sold separately. In certain radio or telegraph system inventions it may be commercially essential to claim the transmitter and the receiver separately. Again, if the invention is a p.c.m. coding technique, one is well advised to have separate claims to the coder and the decoder. A third example is in the field of specialised electrical connectors, where separate claims to the plug and the socket may be commercially desirable. In cases such as these, the commercial need for two aspects being separately claimed may lead to extra expense, since they may turn out legally to relate to two inventions. This would entail the filing of two patent applications if both must be covered separately.

In addition to independent claims, there are usually subordinate claims, each of which starts by quoting the preamble of an earlier claim, and adding some extra feature to it. In almost all cases such a subordinate claim is narrower than the earlier claim to which it refers. If one is reading the claims to assess whether they cover what one wishes to make, it is usually adequate to read only the independent claims. If none of them cover what one wishes to do, it is usually safe to say that one can avoid infringement.

As already stressed, a patent specification is a legal document and is subject to the usual, extremely strict rules which our Courts have established for the interpretation of documents. A claim is, or rather it should be, a specific legal statement of a technical fact, i.e. it expresses in legal terms the scope of the invention for which protection is sought. Therefore, the main claim of a patent specification, is, or should be, the best possible statement in brief of what the invention is. This is one reason for repeating the main claim in the opening paragraphs of the specification as a statement of invention.

The claims usually include a so-called omnibus claim which

relates to an arrangement 'substantially as described and as shown in the drawings', or some such wording. This is in effect a very detailed claim to what is actually described and shown. That such a claim is useful will be clear from one decided case (*Raleigh v. Miller*, 65 R.P.C. 141*) in which the House of Lords held that an omnibus claim was the only valid claim, and that it was infringed. The claims will often include (before the omnibus claim) one or more fairly detailed claims to what the patentee hopes to sell, these often being written as independent claims. As one distinguished patent agent used to say, 'You must have a claim to what you intend to sell'.

If the invention is such that it can be claimed as a method, then it is preferable to have claims to the method, the machine and the product. This is because a claim to a method of doing something can only be infringed by the method when being worked, i.e. only when switched on. By contrast a claim to a machine is infringed (if the claim is sensibly written) whether it is switched on or not. If a patent agent writes a set of claims to a method, a set to a machine, and one or more to a product, he is not merely writing more pages so that he can charge a little more. He is seeking to cover all aspects of the invention. Unfortunately, in many cases a new and inventive method may use known apparatus to produce a known product. This is often the case in a chemical process invention, and in such cases all one can have is 'method' claims, plus 'product by method' claims.

At this point a few words on reading the patent specification are worth-while. Most people, when they read someone else's specification, do so to assess the question of whether it will be a nuisance. In this case it is usual to start with the independent claims. If these are difficult to understand, go to the beginning of the specification and read the opening paragraphs, and possibly look at the pictures. Then try the claims again.

The purpose of the claims about which so much has been said is to define the scope of the monopoly claimed, i.e. to stake out verbally the area from which the author wishes to exclude others.

* *Reports of Patent, Design, Trade Mark and Other Cases,* Volume 65, Patent Office, 1948. At that time the volumes of reports were numbered sequentially from Volume 1, but they are now identified by the year in which they are printed.

Alternatively, instead of excluding others, the patentee can grant licences to use the invention, in return for some financial or other return.

If an article made by a competitor falls within the claims of a patent, then assuming that the claims are valid, the competitor would infringe them. The competitor's article is often similar to what is described in the specification but outside the scope of the claims, which may mean that it was carefully designed to be as close as possible to the article described but to avoid the claims.

The point just made about validity is significant, because patents are granted by the Patent Office without any guarantee of validity. Although the Patent Office searches prior art in an attempt to find anything the same as or similar to what is claimed, there is so much published nowadays that the person who does the search may easily miss some relevant document. Hence, if a patentee finds a competitor infringing his patent, he would do his own investigation into the validity of his patent, and similarly if worried about infringement of the claim of a competitor's patent, the validity of those claims would be investigated.

When a patentee notices that a competitor is infringing the claims of one of his patents and considers that he has a good case in respect of validity of those claims, he usually makes an approach to the infringer. As will be mentioned later, great care must be taken as to how this approach is made. Company policy will usually be the deciding factor as to whether the infringing is to be stopped, or whether to grant a licence. As a last resort, legal action may be called for.

When suing for infringement one can only get damages dating back to the date on which the Patent Office publishes the specification, which may be three and a half years after the original date on which the application was filed. Someone could well start to make an article which would infringe during this period of limbo, but when the publication took place would have to stop infringing (unless he took a licence, in which case the use would no longer infringe). If such use led to an action for infringement, damages could only go back to the publication date. The grant of a patent takes place at least three months after publication (see Chapter 16), and one must wait until grant to sue for infringement.

One is not forced to sue immediately one detects an infringement.

It is possible to wait until the volume of infringement makes it worth the trouble, subject to the Statute of Limitations. This, which does not apply to Scotland, limits the patentee to damages for the six years backwards from the date of commencing the action, or of approaching the infringer.

9

International Arrangements

A number of countries, including all of the important industrial ones, are parties to an International Convention for the Protection of Industrial Property. One of its most important aspects is that if an application for a patent is filed in one of the member countries, then the original applicant or someone deriving a right from him can, when he files in another member country, claim as his priority date the date of filing in the original country. This is subject to the limitation that not more than one year has passed between the two filing dates.

In addition to the International Convention there are a few similar agreements between small groups of countries to similar effect. Thus, although India and Pakistan are not members of the International Convention, Great Britain has agreements with them giving similar rights on a reciprocal basis. Probably, a similar agreement will eventually be made with Bangladesh.

Such agreements have the result that after filing a patent application, the inventor or his company can safely publish the invention, or exploit it, after the filing date, without such publication spoiling his chances of filing patents in other member countries. This is useful to all owners of valuable inventions, but is especially valuable to international organisations.

The 'convention period' for trade marks and registered designs is only six months (see Chapters 12 and 14).

In addition to the International Convention there are a number of other agreements in being or likely to come into being for international co-operation to reduce the cost and labour needs to obtain patent protection in several countries. One such arrangement exists in Scandinavia: Norway, Sweden and Denmark now have almost identical patent systems, and eventually intend to make arrangements for filing what would in effect be a single Scandinavian Patent. What would happen to this if one or more Scandinavian countries enter the Common Market, and the Common Market Patent materialises, is unclear.

A previous European move in the direction of international co-operation is the Strasbourg Convention. Signed in 1963 by eleven countries but so far only ratified by the Republic of Ireland, it is an agreement setting out basic requirements for patentability. That is, it is only concerned with harmonising national laws and not with setting up an international patent. Its implementation would call for changes in most national laws, and the provisions have been taken into account in the recommendations for patent law reform by the Banks Committee.

The establishment of the Common Market gave an impetus to establishing a European Patent, and this will almost certainly be established within the next few years. With Great Britain (presumably) by then in the Common Market, this will be important. The proposals envisage one search for each application, made by one of a group of prescribed searching countries, followed by early publication of the application. Examination of the application, it is suggested, should be deferred and only performed when a fee is paid by the applicant or a third party whose business interests would necessitate him finding out what a patent granted on that application would cover. It is also proposed that if no examination request is received within a prescribed period from publication (unsettled but suggestions vary from two to seven years) then the application lapses. Naturally the application has to be kept alive by paying annual fees. With such a system it is envisaged that all the work of assessing patentability is done by one Patent Office, whereafter a 'bundle' of patents would issue, one for each country prescribed by the applicant when he originally filed. This would operate 'in parallel' with the national systems.

There is also a Patent Co-operation Treaty which embraces the whole world, including the USA and USSR; this is not concerned with setting up an international patent but merely with reducing the cost and labour of getting patents in several countries. It envisages setting up five international searching authorities, and a search by one of these would normally be accepted by all member countries. Where an important patent is involved, with filing called for in, say, 20 or 30 countries, this would save much time and money.

10

Government Security and Crown Rights

Nowadays much research and development is for, and wholly or partly funded by, Government departments. This, of course, means that, depending on the nature of the research contract, the Government will have certain rights under patents arising out of the contract. Another result is that these contracts may lead to inventions which must be covered by the 'blanket' of the Official Secrets Act. Inventions which fall under this 'blanket' can also, of course, come out of non-Government funded work.

Where national security requires it, as in such fields as nuclear science, radars of various kinds, and weaponry, national security may necessitate keeping an invention secret. In such a case the inventor (or his company) can apply for a patent but the application is declared secret, in which case all documents connected with it must be properly locked up and blotting paper and typing carbons, etc., used in connection with it locked up or destroyed. The patent application is dealt with in the usual way as between the applicant and the Patent Office, except that security precautions are observed. When the application is accepted, the documents are not published, but the application goes into a sort of limbo. If subsequently its contents cease to be secret the accepted application may be granted and published. Where such secrecy exists the owner of the invention can, according to the Patents Act, obtain compensation for his inability to exploit his invention. What he usually hopes for is a good Government contract.

At this point the matter of the Crown's right to use a patented invention should be mentioned. The Crown has the right to use any patented invention for Government purposes—this includes the

National Health Service—but this use is subject to compensation for the patentees.

To enable the authorities to decide whether an application for a patent should be declared secret, it is checked on filing at the Patent Office, and if the official doing this checking thinks that secrecy may be needed he issues a secrecy order and sends the papers to the Government department appropriate to its subject matter. During the term of a secrecy order the applicant must, by the terms of the Patents Act and the Official Secrets Act, maintain full security precautions on the invention. Such a secrecy order is continually reviewed and may be cancelled by the authorities at any time.

The Government department who receive the papers of such an application considers them and either confirms or cancels the secrecy order, as it thinks fit. To give time for such considerations a British resident (not limited to British subjects) cannot legally file or cause to be filed an overseas application unless more than six weeks has passed without a secrecy order since British filing occurred, or a secrecy order has been issued and cancelled, or special authority obtained.

The wording 'file or cause to be filed' is necessary, since it means that if the inventor is resident in this country he cannot assign his right to a foreigner to circumvent this provision.

The provision, which really means that if the inventor lives in this country the invention should first be filed here, can be annoying if the inventor has to travel abroad in the course of his work. Where this involves visiting foreign associate companies, the risk of making an invention, or of co-operating with a foreign resident in having an invention, can be high. In such cases to satisfy British law it is desirable to first file in this country.

In the context of this chapter, a person ceases to be normally resident in the United Kingdom if he is abroad for more than six months.

The security provisions discussed above apply also to Registered Designs.

11

The Patent Office

The Patent Office is a department of the Board of Trade, and the information in this chapter is based on the annual reports of the Comptroller-General of Patents, Designs and Trade Marks. The total cost of running the Patent Office in 1967 was £4 376 279, of which salaries, wages, allowances, S.E.T. and superannuation amounted to £2 816 112. Most of the cost was on the patents side, as distinct from the Trade Marks and Designs Branches, both of which are fairly cheap. In 1967 there were one comptroller, three assistant comptrollers and 514 examiners, of whom 15 were temporary.

The number of patent applications has, until recently, been steadily increasing, and has risen from just over 42 000 in 1958 to nearly 62 000 in 1968. In 1971, however, a slight decrease was noted as compared with 1970. In a similar way the number of patents has increased, from just over 18 000 in 1958 to 43 000 in 1968. This increase in the rate at which applications are filed is causing concern, and the Banks Committee has taken it into account in its recommendations. In other countries the increasing rate of filing applications has already lead to alterations in the law.

In recent years less than half the patent applications have commenced with Provisional Specifications, and many of these expired without a Complete Specification being filed. This usually meant that the idea was found either not to work or not to be commercially viable. Many of the applications which are initially filed with Complete Specifications originate from abroad, and in fact a little over half the British patent applications come from abroad.

In 1967 there were only 322 patent applications by women, out of just under 60 000, although there were quite a number made by companies in which one or more inventors are women.

At the end of 1967 there were 209 216 patents in force, which included 46 patents on which extensions beyond the normal terms had been granted. Of those on which the normal period had not expired, only 3172 were in the sixteenth year, so only about 1½ per cent of the patents last the full term. Some of these probably survive because nobody remembers to take a decision to drop them.

According to the Banks Committee, of the patents in force about a third are actually being worked, which is a reasonably high proportion. Many of the others are held in force for speculative reasons or because they cover good alternatives to what the patentee is actually using.

12

Designs

Although the main emphasis of an engineer's work tends to be, as far as industrial property goes, on ideas of an inventive nature, design is also significant, especially in the fields of so-called consumer products. The types of products on which designs registration is significant include radio set cabinets and the shapes of telephone instruments. A recent example of the latter on which designs protection was obtained was the telephone instrument which the Post Office calls the Trimphone, and to which the designers, Standard Telephones and Cables, applied the trade mark Deltaphone. This instrument was selected by the Post Office for development as the 'luxury' telephone, in competition with instruments developed by other companies. Its appearance is now familiar to most of us as it often appears in television plays, and the like. In addition to patents on the circuit and constructional features of this instrument, there are designs registrations on its external features.

Broadly speaking, a design registration covers the external features whereby one recognises the design. These features are identified in the Design Act 1949 as features of shape, configuration and ornament. These words overlap and have sometimes caused argument as to what they really mean. However, an essential point to make is that protection under the Design Act is not for purely functional features, i.e. features which are solely dictated by the function which the article has to perform. In one of the ruling cases on this point, *Stenor v. Whitesides* (65 R.P.C. 1, in the House of Lords), the article in question was a fuse of special shape, the shape being dictated solely by the job the fuse had to do. An action for infringement of this design failed because of this functionality. Even where more than one shape for a device is possible, it will be excluded from protection under this Act, if it is established that its shape is decided solely by the job it has to do (*Amp v. Utilux*). The House of Lords decision on this case is reported in the *Fleet Street Law Reports*, 1971, page 572.

The fact that designs protection is only for non-functional features does not mean that if something is patentable it cannot be registered as a design. In the case of the telephone instrument mentioned above, the arrangement of the handset and the instrument body included patentable features, and the shape (L-shaped) of the handset and the shape of the body were registrable under the Designs Act.

The method of determining whether infringement exists is basically to compare visually the allegedly infringing article with an article covered by the designs registrations, or with the illustrations thereof. If the two articles look confusingly similar, and the design is new at the date of filing it, then there is infringement.

A designs application, like a patent application, should preferably be filed before any publication has occurred, although as in the case of patents, there are a few exceptions which enable publication to take place before filing. Again, these include exhibitions recognised by the Department of Trade and Industry. However, it is still preferable to file designs application before any publication.

Obtaining protection for a design is usually simple and fairly cheap, since all that is needed is to file with the application some illustrations, plus a brief statement of the features believed to be new. The illustrations, which are legally designated representations, should be adequate to illustrate the features of significance for the design. Usually they include side views, plan views, etc., and photographs are often used, in which case they are mounted on stiff paper of the size prescribed by the Designs Rules. In some cases one drawing is adequate, as was the case for a wire structure for drying 'drip-dry' shirts. This was a simple structure of wire, and one perspective view was adequate. In the case of a radio cabinet, one would expect the views to be side (both sides, if different), top plan and front view.

As will be seen from the brief reference to infringement, the scope of protection obtainable is limited, and it is usually safe to say that an infringing article must be closely similar to that shown in the design representation to infringe. Put another way, relatively small alterations will be enough to avoid infringement. However, this form of protection is cheap and for articles of the 'consumer product' type does protect against close copying by a competitor. For an article which either has a limited life, like this year's radio set cabinets, or has obtained public approval, such protection, even though of limited extent, is valuable.

The number of designs applications in recent years in this country has been of the order of seven or eight thousand per year, with a tendency to decrease. These include a fair number of designs for fabrics and wallpapers. Designs applications of overseas origin are usually about one-third of the total, and only about a quarter of these claim a foreign priority date under the International Convention. Thus, unlike patents, the majority of applications are of home production.

At the end of 1967 there were 9175 designs registrations in force, of which 575 were in their third period of five years. In the case of designs, the maximum life is 15 years, for which one pays in three blocks of five years. The fact that a larger percentage of designs remain in force for the full term than patents is probably due to their relative cheapness.

In 1968 there were 7095 designs applications, of which 1752 were of overseas origin, and of these latter only 451 were filed under the International Convention. Note that the period within which filing is possible with benefit of Convention Priority is only six months. There were 5567 designs registrations granted in 1968, i.e. a little over three-quarters of the applications succeeded.

13

Copyright

When dealing with industrial property generally, one has to consider copyright. This can be important in the case of publicity literature and catalogues. Furthermore, drawings are covered by copyright, and if your competitor makes a so-called 'Chinese copy' of your device, he may well infringe copyright of the drawings. This is based on the theory that to produce his device he needs drawings, which are produced by copying (in effect) from your device. The following statement made by Mr. Justice Graham, in *Merchant Adventurers Ltd. v. M. Grew and Co. Ltd.*, reported in *Fleet Street Law Reports*, 1971, p. 233 *et seq.*, is of interest:

'There is infringement of drawings by a three-dimensional reproduction of these drawings if they are sufficiently clear for a man of reasonable and average intelligence to be able to understand them from an inspection of them to such a degree that he would be able to visualise in his mind what a three-dimensional object, if made from them, would look like'.

Any document which is produced automatically acquires copyright without the need for registration, in which respect our system differs from that of some foreign countries. Copyright in this country normally lasts until 50 years after the death of the author. The protection given is therefore long-lived, especially if the author lives as long as Bertrand Russell. However, copyright is only infringed by actual copying, i.e. if the second document can be shown to be of independent origin, then there is no infringement. This point is strongly made in a standard British textbook on copyright* with the following statement:

* Copinger, W. A., and James, F. E. S., *Law of Copyright*, Sweet and Maxwell, 1965.

'If it could be shown that two precisely similar works were in fact produced independently of one another, the author of the work that was published first would have no right to restrain publication by the other author of that author's independent and original work'.

This point is worth emphasis as it underlines the essential difference between protection by the Copyright Act and by the Patents Act and Designs Act. In the cases of patents and designs, if the later product is within the scope of the protection validly given by the patent or the design, then independence of origin is no answer to a charge of infringement.

If one literary work is 'in copyright', the author's permission is usually necessary if one wishes to publish a translation of it into another language. Any translation thus produced would be covered by its own copyright.

Between designs registration and copyright there is somewhat ill-defined overlap, and also there were cases where neither gave adequate protection. To remedy this, the Designs Copyright Act 1968 was enacted. Before this, there was no protection for an industrial design unless it was actually registered under the Registered Designs Act 1949. The position now is that copyright protection under the Copyright Act 1956, as amended by the Designs Copyright Act 1968, arises immediately on production of the design, and is given to industrial designs which are also works of art (e.g., a mass-produced statuette) for 15 years as if they were registered under the Registered Designs Act. There is no need to register under the Designs Act, but such registrations for some articles is not prevented. Where both Acts apply, one therefore has two strings to one's bow in dealing with infringers.

An important gap in the protection available nowadays is that novel type fonts cannot be validly protected by any of the various systems now in existence. This point is quite significant to the printing trade, and among reputable printers there is an understanding that one does not copy the other man's type font without his permission. Hence protection exists to an extent limited by the printers to which this purely voluntary understanding applies. There are also printers who do not know about this, however, or who do not feel inhibited by it if they do.

Recent developments such as special type fonts for automatic machine reading or optical display devices can be commercially important, and should be legally protectable. Inquiries are now in progress to ascertain the need for such protection and also (it is hoped) to devise some scheme to achieve this.

This chapter does not by any means exhaust the subject of copyright, which even the experts find rather difficult. In fact, the author has heard it said that in this country there is only one firm of solicitors who fully understand the present state of the law on copyright!

14

Trade Marks

Trade marks do not usually need to be considered by engineers, at least in the early or 'design' stage of producing a new product. However, eventually if one is in the field of consumer products one must consider them. The subject is a big one, and for the present purpose the main thing is that a trade mark, to be registrable under the provisions of the Trade Marks Act 1938, should be a word, device, or other symbol which is distinctive of the owner of the mark but does not describe the goods to which it relates. If the mark does describe the goods to which it relates, then it could be suitable for use by anybody making such goods. Therefore, it would not serve the fundamental purpose of a trade mark which is to indicate a connection in the course of trade between the goods to which the trade mark is applied and the person who is making or selling them.

One can use a trade mark for one's goods which is quite unregistrable under our legislation, but such a mark is weaker than one which is registered. This is because if one has a registered trade mark and someone else copies it or uses a mark which is confusingly similar to it, one can sue him for infringement of the trade mark. To prove this, one does not have to prove that someone was deceived. By contrast with this, if one's mark is unregistered and someone pirates it, one has to proceed for passing off. To succeed in such an action one must find independent witnesses (i.e. not one's secretary or wife!) who are willing to swear that they have been deceived. As this can make the witnesses look silly, they are not always easy to come by.

Although a word (or other mark) used within a company to describe an article, possibly because of some descriptive relation, is ideal for internal purposes to describe the product, it may be unsuitable for registration as a trade mark. If a trade mark is being considered, expert advice should be taken to decide whether the mark is registrable, and also to see whether or not someone else has registered it. This should preferably be done before any money has been spent on publicising it.

In considering a word or other mark's suitability for registration, remember that certain things are quite unregistrable, as can be found in the Trade Marks Act. The main things to avoid are:

(1) A common surname.
(2) A reasonably well known geographical location.
(3) A word directly descriptive of the character or quality of the goods.
(4) A word which is misleading as applied to the goods.

Under item (2), it was held many years ago that LIVERPOOL could not be registered for electrical cables because Liverpool is a well known city. This was so, in spite of the fact that Liverpool Cables was then well known, and to anyone 'in the trade' Liverpool was distinctive of their goods. (Liverpool Electric Cable Co.'s application for Trade Mark, 46 R.P.C. 99.)

In a similar recent case (1971) Waterford Glass Limited tried to register WATERFORD for glass: although it was said to be in fact distinctive of their goods, the fact that it is a place in Ireland prevented them from getting it registered.

Foreign markets, if any, should be taken into account when selecting a trade mark. It is not economic to find that as your British mark is unsuitable in, say, France, you need a new mark for that country. It is possible that a mark which is quite respectable here may, in some foreign language, be vulgar or offend against a local religion. Hence, the choice of a trade mark can be difficult.

In this country there are two types of registered trade mark, that registered in Part A and that in Part B of the register. The Act sets out legal distinctions between them, the effect of which is that a Part A mark is stronger, and thus gives better protection to its owner, than a Part B mark. As a corollary to this, a mark not registrable in Part A may well be registrable in Part B.

Another form of trade mark is called a Certification Mark. This is registered by a trade association and specifies standards to which the goods to which it is applied must conform. Such marks are the Harris Tweed mark and the Kitemark of the British Standards Institution. The right to use such a mark is granted to a manufacturer who meets standards set by the owner of the mark.

A registered trade mark, as long as it is valid, can be kept in force for ever by payment of the renewal fees which fall due every seven years. As is obvious, especially to housewives, a good trade mark is valuable. One only has to think of such well known marks as HOVIS and KODAK. A good trade mark in a more technological field is PENTACONTA, registered in various countries by companies of the ITT family for telephone exchanges and related equipment.

The owner of a trade mark should take care not to mis-use it, or allow others to do so. Any unauthorised use of a registered trade mark in connection with the goods which it covers infringes it: for instance, to sell bread and to be so unwise as to state in writing that it is better than HOVIS would almost certainly infringe that trade mark. Any unauthorised use of a trade mark should be challenged by its owner. This may sometimes lead to conflict with the publicity department, but it is legally desirable to keep a trade mark valid.

It is especially desirable to make sure that a registered trade mark does not become the accepted technical word for a product, since it will then no longer be valid because it will fail to distinguish your goods from someone else's. In publicity literature, one should indicate (however discreetly) that one's registered trade mark is, in fact, a registered trade mark. LINOLEUM and KLYSTRON (a special sort of radio valve) were originally registered trade marks, but they became the usual words for the goods by general usage. In the case of KLYSTRON, the main cause was use by armed forces in Word War II.

A trade mark should always be used in such a way as to make clear that it is a registered trade mark. It should at least begin with a capital letter, and should have 'registered trade mark' in brackets, or as a footnote, or as a special note in a brochure. In keeping with this, it is well to respect someone else's mark for two reasons. Firstly, if you do he will probably respect yours in return, secondly, if you don't he will attack you! The publicity department may well argue that identifying a trade mark as such is 'too legal for good publicity', but a good trade mark is worth a little trouble if one can thereby keep it in force.

Trade mark applications are tending to increase in number, and now run in this country at about 17 000 per annum, in most of which the applicants apply for Part A registration. About two-thirds succeed, although one third of the successful ones end up in Part B. This includes many which are tried in Part A, but are eventually allowed in Part B. If one uses a Part B mark for many years until it is well established, one may then be able to register it in Part A. Similarly, if one is forced to use an unregistered mark one may by long use establish it as distinctive of one's goods. If so (and if it is not excluded in the manner mentioned above) one may then be able to register it. However, it is preferable if possible to use a registrable trade mark from the start.

15

Unusual and Interesting Inventions

In the past a number of industries have been founded on the basis of a good invention well patented, and examples of this include the early wireless industry, where Marconi's early patents were significant. Another important early invention was that of the Wright brothers relating to control of an aeroplane by warping the wing. This meant that the wingtip was physically bent, but the United States patent was well written and covered ailerons, which although superior mechanically were functionally the same.

Other valuable inventions include the technique of colour filming which used to be called Technicolour, and which for many years was the only colour film technique ever seen in the cinema, and one by Georges Valensi, described in British Patent No. 524443 on colour television. This was applied for before World War II, but owing to extensions of the term, lasted long enough ($32\frac{1}{2}$ years!) to cover the sale of colour television sets. The patent was owned in this country by EMI, who sued for patent infringement, and won in the court of first instance. They lost in the Court of Appeal, but had the appeal failed EMI would have been in the position to collect royalties from the entire British television industry.

Another example of an important patent is British Patent No. 639178 on printed circuit boards, which has now expired. This was owned in this country by a company called Technograph, who issued writs against a number of infringers, and the action against one of them, Mills and Rockley, came to court. The patentees won after appeals to the House of Lords (1969 R.P.C. 398) in respect of one of the processes much used today, and are now collecting royalties from other infringers.

We can now consider a few of the more comic or peculiar inventions on which patents have been granted. British Patent No. 122080

(Wilson), invented at the end of World War I, related to the disposal of enemy aircraft by what could be called airborne bayonet charges. Claim 1 of the patent read as follows:

'For aircraft a weapon consisting of an adjustable blade fitted to the machine in such manner that the said aircraft may be caused directly to charge an enemy machine and disable or cut it asunder'.

One wonders what Walt Disney would have made of this! Curious though it looked, it was in a way a fore-runner of an idea proposed during World War II by the distinguished American aircraft designer Jack Northrop. This was a small jet fighter with reinforced wing leading edge, intended for ramming enemy aircraft.

Going back to the 'pre-history' of patents, i.e. before the present requirement for claims to define the invention, we have Patent No. 418 of A.D. 1718, in which Mr. Puckle described a portable gun, one of the functions of which was to shoot square bullets against Turks and round bullets against Christians.

British Patent No. 4413 of 1903, of a Mr. Whitney, who described himself as a lawyer and a scientist and a citizen of the United States of America, was a method of collecting electricity from the ether for use on earth. He collected his electricity by a cable carried up by airships or fired up by a cannon. How it stayed up in the latter case was not explained.

In those early days the present practice of the Patent Office of examining each application was either non-existent or in a very rudimentary stage. This really explains some of the oddities of the times.

Another bright idea, British Patent No. 9329 of 1904 (Fitzmaurice), was floating soap for advertising purposes. The soap bore an inscription, or was transparent with an inscription inside it.

British Patent No. 360253 (J. Lyons and Co.) was a practical idea for a multi-spout teapot. Claim 1 of the patent gives a clear and unambiguous statement of the invention:

'A teapot or like portable pouring vessel comprising a plurality of spouts from which the contained liquid may be poured simultaneously'.

The author remembers seeing such a teapot used by the organisers of a whist drive so that they could quickly pour the 30 or 40 cups of tea at half-time.

Other peculiar subjects for which patents have been granted include apparatus for preventing snoring, an arrangement whereby a horse-driven carriage had the carriage ahead of the horse so that a machine-gun could be mounted in the front of the carriage, and an apparatus for detecting buried corpses.

An invention of particular interest in the light it casts on the habits of foreigners was British Patent No. 3069 of 1910. The inventor was Dr. Rudolf Lux, who was a judge resident in Silesia. The invention was a contraption for protecting judges against assassination, and the patent included the following claim:

'A device for protecting members of a court assembly against assassination comprising a bullet-proof plate hingedly connected to the head piece of the court table, said plate being normally maintained in horizontal position and adapted when released to be raised by means of a slidable spring-actuated rod into vertical protecting position, means being provided at each seat of the table for releasing the protective plate by pulling a knob substantially as set forth'.

The last four words 'substantially as set forth', or something similar, were in those days much used in claims, but have now disappeared (except from omnibus claims).

Another oddity which could lead to frivolous comments was British Patent No. 5785 of 1907, in which the inventor, a Mr. Mumford, had invented trousers with 'flies' at the back as well as at the front.

The above selection, which is by no means exhaustive, shows that patent work has its lighter moments. One canard which keeps cropping up is the perpetual motion machine. People still dream of making such a machine, but if they try to patent it as a perpetual motion machine the Patent Office will reject it as being contrary to established natural principles.

An interesting field of invention which patents agents have to consider these days is computer programs. These at present can be covered by copyright, but as explained in Chapter 13, this only protects the first inventor against actual copying, as distinct from a later devising of the same idea. Further, the scope of copyright protection is relatively narrow as it does not go to the fundamental basis of the idea involved. Hence, attempts are made to get patents for a computer which is programmed in accordance with the new program. As a result of some recent decisions in the Patents Appeal Tribunal by the late Mr. Justice Lloyd-Jacob and Mr. Justice Graham, one can patent a computer program if the specification is carefully written and the claims suitably worded. Such claims are claims to a computer, any old computer, programmed in a special way, and in enforcing them one might come up against the principle that the other person, having bought a conventional computer, is entitled to use it in any reasonable way. Another argument is that since a program is analogous to a law of mathematics, and since such laws are inherently unpatentable, then so too is a program.

In view of these doubts, if the program is valuable, all one can do is either keep it a secret, with all the risks that involves, or patent it and hope that the patent will be enforceable. Such a patent would at least prevent anyone else from subsequently getting any form of legal protection for your program.

A recent example of a patent for a computer program is British Patent No. 1176801 (J. F. Canguilhelm), which is an extremely learned treatise on dimensional synthesis, a simulation technique applicable to the assessment of business systems, defence plans, and the like. The first claim of this patent begins:

'A computer when programmed to perform dimensional synthesis on a plurality of sets of values representing elements in the form of points or vectors in a multi-dimensional space, the synthesis proceeding according to one or more of the following sub-programmes. . .'.

Then follow six sub-programs, each of which is a mathematical expression. No 'hardware' is described in the specification, so the claim is merely to a computer—any old computer—when programmed in this way. Corresponding patents have been filed in several other countries, including the USA.

The question of how to get valid and useful protection for a computer program is under active discussion, and it would seem that a program is only worth more than copyright in a few cases. These would include automatic control programs, computer-assisted design programs, and programs for controlling telephone exchanges. All of these are 'industrialised result' cases. Where one has a good 'program' idea in such a field it is worth asking the patent agent if anything in it is patentable.

16

Obtaining a Patent

In this chapter it is assumed that the inventor, or his employer, are employing a patent agent to obtain the grant of a patent for the invention. As already mentioned, the inventor himself can deal with his own patent application, but it is seldom advisable to do so since the patent specification is both a legal and a technical document.

As a first step the inventor should provide his patent agent with full information concerning his invention. The patent agent will then, on the basis of this information, write a patent specification which should, ideally, be a good engineering description and also a good legal document. The description supplied to the patent agent should, preferably, include the principle, object, range, field of application and full description of one or more arrangements using the invention. It should also include any relevant prior art, i.e. technical journals, books or patents of which the inventor is aware, which should be explained to the patent agent adequately. It is rare indeed that a patent agent will receive such a perfect description from an engineer: in fact his 'briefing' is usually exceedingly inadequate! The amount of description which the patent agent requires will, of course, vary from agent to agent and one can usually assume that the patent agent has enough sense to ask for more information if he needs it.

The next step is to decide whether to file a Provisional Specification or a Complete Specification, and it will be assumed for the moment that a Provisional Specification is chosen. Its main purpose is to establish a date for the invention described in it. Theoretically a Provisional Specification can give a description of the general nature of the invention, and may include one or more examples of embodiments of the invention and indications of possible develop-

51

ments. There are, however, no claims in the Provisional Specification, i.e. no statements whose purpose is to define the limits of the monopoly to be given by the patent (when granted). However, it is a good idea to include one or more paragraphs which foreshadow possible claims for use when the Complete Specification is to be written. To justify possible broad claims, it is desirable to put as much as possible into the Provisional Specification. It is also desirable for other legal reasons that where an invention can be illustrated by a drawing, the Provisional Specification should include a drawing. This is, indeed, almost essential if one later files in the USA and claims as priority date the date of the British Provisional Specification.

Two copies of the Provisional Specification, together with the official application form, stamped £1, are lodged at the Patent Office as soon as possible, and the application is then given a number by the Patent Office. This is a number ranging (nowadays) between 1 and about 60 000, followed by an oblique stroke and the year number, for instance 37322/73. The application forms should be signed by the applicant, or by somebody holding a power of attorney on his behalf, should name the inventor and should be signed by the inventor (or alternatively be accompanied by a document which the inventor signs whereby he gives his consent to the application).

The purpose of the Provisional Specification is to establish the date on which the invention was made, and for this a general description is in theory sufficient. However, since a claim to a specific embodiment might, if that embodiment is mentioned in the Complete but not in the Provisional Specification, have as its effective priority date the date of the Complete Specification, it is highly desirable that as much as is known should appear in the Provisional Specification. This point about 'multiple priorities' is significant where, as soon as one Provisional Specification is filed, the invention is publicly used, e.g. by sale, or if it is described in some publication such as a paper read by the inventor. If one carelessly uses in public or otherwise publishes an embodiment of the invention not in the Provisional Specification before the Complete Specification is filed, any claim specific to that embodiment would fail due to one's own public use or other publication.

When the application is initially made, it is possible to file a Complete Specification, in which case the official fees consist of £1, payable on the application form, plus £22 payable for the Complete Specification. It should be noted that this £22 fee is also payable when one files the Complete Specification, if one originally filed a Provisional Specification. The advantage of filing a Provisional Specification is that, as it is cheaper than a Complete, it is possible to file a Provisional relating to an invention while it is still in a relatively embryonic stage, in which case the application can be abandoned without much loss if the invention fails. A further advantage is that the period of effective protection for the invention is greater, as the term of the patent commences with the day on which one files the Complete Specification. By comparison, it will be appreciated that if one files a Complete Specification in the first instance, there is some saving of office work as compared with filing a Provisional followed by a Complete Specification, and also in general some saving in expense.

The Complete Specification should be filed, with its prescribed fee, within one year of filing the Provisional Specification, but this period can be extended on payment of a fine by up to three months. If no Complete Specification is filed, the application is automatically abandoned.

The Provisional Specification was originally intended to be a cheap and simple way for an inventor to establish a priority date for the invention, the year before he filed a Complete Specification, giving him time (a) to complete his development of it in a practical form and (b) to interest a possible purchaser or licensee of the invention. Nowadays these purposes have to some extent receded into the background because most inventors are employee-inventors.

The Complete Specification gives a full and detailed account of the invention and the best method of performing it for which the applicant is entitled to claim protection. It ends with a set of claims which, as already stated, define the invention. It is these claims which by their scope and by the care with which they are drafted, define the extent of protection given by the patent and hence the value of that patent. In general when interpreting these claims they should be read on their own, but one can refer to the specification i.e. one can use the specification as a 'dictionary' for the claims if they are not immediately clear.

It is this requirement for claims which is a major difficulty in writing a Complete Specification. A claim should be as wide as possible so that all sorts of arrangements using the inventor's idea are covered, but it must not be so wide as to be vague, or to cover something which does not work, or to cover something which is old or obvious. To draw a line between these conflicting requirements is difficult, which is why one is well advised to go to a specialist when filing a patent application. A special difficulty arises where the real invention is the appreciation that a certain problem exists. In such cases the object of the invention and the means for satisfying that object are so closely associated that it is almost impossible to attain the same object without infringing the patent, but in other—in most—cases it is possible to attain the same object by many different methods, each of which may itself be patentable.

The Complete Specification must, in its descriptive portion, include enough detail to enable one 'skilled in the art' to work the invention without any further exercise of invention on his own part. It is permissible to complete the description by reference to other published documents, and it is also possible to assume such elementary facts as one skilled in the art would know. At this point it is worth mentioning that the extent of knowledge of one 'skilled in the art' will vary between different arts, and furthermore that where an invention lies in a comparatively new field of the art a more detailed description may be necessary than would be the case in one of the more old and long established arts. The fact that tests may be needed to put the invention into practice does not usually mean that a description is inadequate. For example, in inventions on electrical circuits it is usually unnecessary to give specific values for components unless they are critical.

Methods of drafting Complete Specifications vary widely among different patent agents, but in all cases a good Complete Specification should be an adequate technical description which is susceptible of only one meaning. This one meaning should be the meaning which the writer intended. Where the writer has been either careless or unlucky, some passage which he has included and which was perfectly clear to him may be given a wrong meaning (or even no meaning!) by the Court, with possibly disastrous results to the validity of the patent. At this point it should be mentioned that it is a principle of interpretation of legal documents that where a document can be given one meaning, it will be given that meaning even if the writer establishes that he never intended it. Evidence of the writer's intention is, in general, only relevant where a document can, on its face, be given two or more meanings. This again emphasises the necessity for great care in drafting a patent specification. For this reason it is usual for a patent agent to obtain his client's (or the inventor's, in the case of a company patent department) approval both of a Provisional and a Complete Specification.

This does at least mean that the descriptions are (or should be) checked for technical accuracy.

It also means that if the patent agent is reproached for some technical bug in the specification at some later date he can successfully transfer the blame to his client (or the inventor). For this reason the vetting of a draft specification submitted by the patent agent is an important and often vital matter since not only can it be crucial to the validity of the patent, but if badly done the client or inventor (legally) has only himself to blame. Thus, the client or inventor must co-operate with the patent agent!

When the Complete Specification has been filed at the Patent Office, it is allotted to one of the Examiners who does a detailed examination. He investigates whether it is correctly described, whether it relates to a single invention (if the claims included related to two inventions or more, the applicant will be required either to modify his claims so they relate only to one invention, or to file separate applications for the other inventions), whether it is lawful and moral, and whether it has been anticipated, i.e. already described, by any previous specification filed in Britain during the last 50 years.

Considering these points in more detail, where the claims relate to two or more inventions it is possible to file separate applications for the inventions removed from the specification, and the date carried by these applications would be the same as the date carried by the original application.

It was stated above that the search relates to British patent specifications: in fact, the Examiner is entitled to quote as a possible anticipation any document published in this country, 'published' meaning available to the public as of right. However, a book or magazine in the library of a learned society such as the I.E.E. and only available as of right to members of the society is regarded as being published because it can be consulted by a substantial proportion of the experts in that society's field. There are certain specific limitations to this entitlement to quote published documents: for instance, a patent specification or an official abridgement thereof is only counted as an anticipation if it was published in this country not more than 50 years before the application date (or the priority date in the case of an application filed under the International Convention) of the application being examined.

When the Examiner has completed his examination he either accepts the application without issuing any further communication (this happens occasionally) or he issues an Official Letter which is sent to the applicant via his patent agent. The latter can then, on behalf of his client, either abandon the application, or amend to avoid the objections raised by the Examiner, or if he considers them to be unjustified, he can contest the Examiner's opinion. If the patent agent and the Examiner are unable to agree, a Hearing can be arranged in which the matter is argued before a Hearing Officer. A Hearing is, in effect, a judicial proceeding, although it is usually conducted in a relatively informal manner, and in most cases the decision is given verbally. If the patent agent, or rather his client, considers that a rejection by the Hearing Officer is unjustified, he has the right to appeal to the Patents Appeal Tribunal. This consists of a High Court Judge, normally one of the special judges appointed to deal with patent and related matters. At the time of writing, this means either Mr. Justice Graham or Mr. Justice Whitford.

When the Examiner has been finally satisfied, or the applicant has won his appeal, the application is accepted by the Patent Office and the Complete Specification printed. Although nowadays the Provisional Specification is not printed, it becomes open to inspection and can be studied at the Patent Office by any member of the public. When the application has been accepted, its acceptance is advertised in the weekely Official Journal and it is allotted a seven-figure number. The Official Journal also states when printed copies of the Complete Specification become available, nowadays at 25p each.

The application should be in order for acceptance, i.e. the Patent Office should be satisfied, not more than two years, six months after the filing of the Complete Specification, although this period can be extended by up to three months, though there is a fee for this privilege.

The Patent Office recognises that its employees are not infallible. Indeed, the Examiner cannot be, since the range of his search is restricted; this is why a patent is not granted until three months after it has been printed. At any time during these three months any interested party (this will usually be some person other than the applicant but working in the same industry) can oppose the grant of a patent on one or more of a number of grounds.

These grounds include the main grounds on which the Examiner can reject the application, and which are discussed above, and they also (since 1949) include the ground that the alleged invention is clearly obvious as compared with what was known before the date of the patent. This extends the powers of the Patent Office, but is somewhat limited, as in order successfully to establish that a claim of a patent is obvious before the Patent Office, one has to produce stronger proof of obviousness than is necessary before the Courts. The difficulty here, as when one argues obviousness before the Courts, is that obviousness is inevitably a matter of opinion. The usual grounds on which the application is opposed are that the alleged invention is not novel, in which case one always quotes the documents in support of lack of novelty as supplying evidence of obviousness in case they are not held to destroy the claim completely on the ground of lack of novelty.

One can also oppose the grant of a patent on the ground that the inventions claimed in any one of the application's claims have been claimed in a patent specification of earlier date, not published at the time when the opposed application was made. This ground of prior claiming can also be used by the Examiner during the application stage. However, the application cannot at that stage be rejected solely on this ground. If the applicant does not wish to amend his claims during the prosecution of the application prior to acceptance to remove conflict, he can allow the Patent Office to insert a specific reference to the patent in question. However, it is possible for an application to be rejected on this ground as a result of an opposition.

If the application is opposed, then proceedings of a judicial nature occur, firstly before a Hearing Officer at the Patent Office, and secondly, if (as usually happens) the unsuccessful party in the opposition is not satisfied, on appeal to the Patents Appeal Tribunal.

If no opposition is encountered, or if an opposition has been successfully dealt with, the patent is granted. This, as usual, necessitates payment of a fee. Details of the grant are entered in the Official Register of Patents, which contains details of all patents in this country.

It is possible to apply to the Patent Office for a patent to be revoked during a period of one year from its grant, on the same grounds as one could oppose the grant during the previous three months. This procedure, which is known as 'belated opposition', is usually used where an interested party has forgotten to oppose the grant. However, it is not permitted for the same person to oppose before grant and also to apply for revocation to the Patent Office after the grant.

An application to the Court for revocation is, however, another matter altogether, and can be done at any time after the grant, and on a wider range of ground than can opposition or 'belated opposition'. In this case there is right of appeal to the Court of Appeal, and in some cases to the House of Lords.

17

Patentee's Obligations

The monopoly given by the grant of Letters Patent normally lasts for 16 years from the date of filing the Complete Specification, provided that the necessary renewal fees are paid. The sealing fee paid when the patent is granted keeps the patent in force for five years after the filing of the Complete Specification. This comes quite quickly after grant as the acceptance can take up to two and three-quarter years, plus three months more before grant (possibly plus a further hiatus if there is an opposition). Where there is an opposition the renewal fees can theoretically fall due before grant, but any such fees falling due would not be payable until after grant.

The first renewal fee is now £13 and the annual fees increase so that for 16 years one has to pay over £300 in renewal fees. This underlines the importance of a type of patent not mentioned earlier: the Patent of Addition. Where one has an invention which is an improvement on or modification of the invention of one's earlier patent, the new application can be for a Patent of Addition to the earlier patent. Such a patent pays normal fees up to grant, but no renewal fees. However, it dies with the parent patent. If one converts the Patent of Addition to an independent patent it is necessary to pay renewal fees for the period extending back to the date of the Patent of Addition, but the expiry date of the newly independent patent is unchanged. Hence, it is necessary to balance the saving in renewal fees against the limitation in the term. Patents of Addition are quite useful for the actual 'production version' of a device covered by an earlier patent.

There are two circumstances in which the term of a patent can be extended. In the first case an extension of five or ten years can be obtained if the patent is for a valuable invention for which there has been inadequate remuneration. This needs an often costly appli-

cation to the High Court, which may be opposed. As really out-
standing inventions are rare, so too are such applications. However,
a recent one was Sir Frank Whittle's patent for the by-pass jet
engine. Another was the Valensi Patent No. 524443 on colour television (see Chapter 15).

The other ground for extension is where the patentee has been
prevented from exploiting the invention by a state of war between
Her Majesty and a foreign power. Here the term granted depends
on an estimate of how much time has been lost. Application is to the
Court or, more usually, to the Patent Office.

A war loss extension and an extension for inadequate remuneration
can be granted successively on the same patent (e.g. in the Valensi
Patent), and also two extensions for inadequate remuneration are
possible. Extension of term granted as above, whether for inadequate
remuneration or for war loss are not subject to renewal fees. Hence,
eventually one is actually given something—after the trouble and
expense of applying for an extension and possibly fighting oppositions
thereto.

The main obligation of a patentee is that he does not abuse his
monopoly rights. That is, he must not misuse his patent. This is
a topic about which there has been much ill-informed gossip in
the press but it is, in fact, not very common for a patentee to abuse
his monopoly. The Act says that public demand for the invention
should be satisfied by manufacture here (in preference to manufacture abroad with import). If the market is not so satisfied then
an 'aggrieved person' can seek relief by applying for a compulsory
licence.

The Act also says that no trade or industry should be unfairly
prejudiced by conditions imposed by the patentee. One practice
frowned on is the grant of a licence with tying clauses, e.g. making
the grant conditional on obtaining supplies from a specified source,
usually the licensor. The existence of such a licence granted by
the patentee to a person A is a good defence to an infringement
action by the same patentee against a person B unless the patentee
shows that he also offered a person A a licence on reasonable terms
but without tying clauses. This is usually done by offering two
licences at once, one with and one without tying clauses, with the
former charging lower royalty. As long as both royalties are
'reasonable' this is acceptable.

If the charge of abuse of monopolies is upheld the patent may
be ordered to be endorsed 'Licences of Right' on the Register of

Patents. This means that the patentee must grant licences to applicants on 'reasonable terms', any disputes being settled in the Court. A patentee can voluntarily have his patent so endorsed, the point being that it halves the renewal fees. However, it constitutes an obligation to grant a licence 'on reasonable terms' to anyone who seeks a licence.

An interesting ground for an application for a compulsory licence is that an export market is not being satisfied as a result of the patentee's unwillingness to grant a licence. It would seem in this case that a compulsory licence might be ordered even if the patentee's factory is exploiting the patent at full blast, although there appears to be no case law on this point!

One point about patents endorsed 'Licences of Right' is that when the endorsement produces an application for a licence, the endorsement is often cancelled, the reason being that many people wish to be sole or exclusive licensee.

It is specifically forbidden in the Patents Act to threaten an action for infringement. Hence, when one notices an infringement a brief letter mentioning the infringing product and mentioning one's own patent as being of interest is acceptable. If one notices infringement, one should take legal advice before (and not after) approaching the infringer. Obviously, unjustified threatening is also a common law offence.

If one does threaten an action for infringement, this gives the other man the right to sue for threats. Proof of infringement is usually an adequate defence. However, most lawyers feel that it is tactically preferable (where possible) to be the plaintiff rather than the defendant. Hence the need to take advice.

18

Patentee's Rights

This and the preceding chapter on obligations to some extent overlap.

The rights of the patentee include the right, by legal action if necessary, to prevent unauthorised persons from using his invention, whether the infringement is by manufacture in this country or due to import. A 'loophole' here is that use of a patented article in a ship or aircraft temporarily within British jurisdiction is not an infringement provided that it is only for the use of that ship or aircraft. In such cases the guiding principle seems to be the country of registration of the ship or aircraft.

If an infringement action succeeds, it can result in an injunction against further infringement, delivery up or sometimes destruction of infringing goods still in the possession of the infringer, or payment of damages.

The usual defence against an infringement action, apart from the obvious one of denying infringement, is to attack the validity of the patent. Hence, the defendant usually counter-claims that the patent is invalid, and if he succeeds the plaintiff fails as the invalidity, if established, is retrospective in effect. The counter-claim may invalidate enough of the patent to kill the infringement action but leave some useful residue of the patent. The chance of a counter-claim for revocation makes an infringement proceedings somewhat costly. Another defence mentioned earlier is to show abuse of monopoly by the patentee in his granting of an improper licence, i.e. one with tying clauses in it.

The patentee can assign the whole or a part of his patent rights, or sell the rights, or grant licences to work the invention, either with or without payment of royalties. There are various forms of licence, so that one can grant an exclusive licence, which excludes the patentee from working the invention, or a sole licence, which excludes everybody other than that licensee and the patentee, or a non-exclusive licence. There can be several exclusive licences under one patent: if the claims relate to an electronic circuit, there could be exclusive licences for the application of the invention to transistors, and for its application to vacuum tubes. The same applies to sole licences. If an infringement occurs and the patentee will not sue, an exclusive licensee can sue on his own behalf.

The patentee can drop his patent, usually merely by not paying the renewal fee, although mechanism exists for relinquishing the patent. The question of maintaining a patent obviously depends on the value of the invention and the use made of it. If the patent has a wide claim but is not in use, maintenance may be warranted on speculative grounds or merely to annoy competitors! Again, if the claim is narrow but the invention is in extensive use, retention may be useful to restrict competition.

Should a patent lapse inadvertently, an application can be made to revive it, but unless the application is made quickly the process is complex and one has to give good reasons to support a claim of inadvertence but not pure carelessness. There are safe-guards here for people who commenced to use the invention when the patent lapsed.

As mentioned in Chapter 17, extension of the term is possible, either on the ground of inadequate remuneration for a valuable invention, or on the basis of loss of opportunity to work the invention due to a state of war.

19

Patent Agents

In earlier chapters patent agents have frequently been mentioned. Ideally a patent agent is both a specialist lawyer and an engineer, so that he (or she) can understand both the legalities and the technicalities of the invention.

The usual line of recruitment is direct from university after a science degree, although some enter the profession via industry. The latter route is better than the former, since that way the recruit will have acquired some practical engineering knowledge. In either case, on entering the profession, our recruit starts with relatively routine work, often including doing searches, to give him some understanding of this routine. He will also prepare specifications, and reply to official letters from the Patent Office, but in these cases under careful supervision. Often this supervision is difficult, because there are often several ways to do a job, all equally good.

In this country, unlike the situation in some others, a patent agent has to pass qualifying examinations conducted by our professional body, the Chartered Institute of Patent Agents. The regulations for these examinations specify preliminary, intermediate and final stages.

The preliminaries are formed by certain examinations conducted by other bodies such as Higher National Certificate and University degrees, the nature of the preliminary qualification defining the time at which the young aspirant can take the intermediate examination.

The intermediate examination which, if one has a science degree from a British university, can be taken after a year of working for a patent agent, includes a variety of papers. These include British Patent Law, British Trade Marks and Registered Designs Law and foreign patent law. The latter is desirable as a patent agent often has to give advice on the situation overseas. Usually he would, in major matters, check with experts in the countries concerned. There are also papers on patent agent's practice, which seek to simulate the sort of day-to-day work of a patent agent. Under this heading there are writing specifications, usually on relatively simple inventions. Finally, there is a 'Manufactures' paper which attempts to test the candidate's general technical knowledge and his ability to write a good technical description. This paper, when the author took it in 1951, included 20 questions, of which he had to answer 10, in a three-hour session. Examples of the questions, from the 1970 paper:

    (a) Describe the manufacture of a soft soap.
    (b) Illustrate and describe the circuitry for a small 3-phase motor with a drum controller for motor reversing.

There is also a very popular question at the end:

    (c) Illustrate and describe any manufacture not the subject of any of the previous questions.

If the candidate passes this examination he can take the Final Examination two years later. This includes papers on the preparation of specifications from the sort of information a client would give, patent agent's practice papers and interpretation and criticism papers. The last named are usually the most difficult: the candidate receives a description of what a client wants to make and sell, and a patent specification and some specimens of prior art. He is asked to advise his client as to whether the patent is valid in the light of the prior art he has been given, and also whether the client's device infringes that patent. The questions are difficult to prepare, difficult to answer and also difficult to mark. They seek to test the candidate's ability to reason intelligently from a set of facts and also his ability to express himself on paper.

The Finals papers are in groups and a candidate can pass all groups at once, or pass them group by group. This grouping is relatively new, but has not made a tremendous difference to the pass percentages. When the Finals have been safely passed, the successful candidate can have his name entered on the Register of Patent Agents, for which there is an annual fee, and can put up his brass plate! He can also apply for election as a Fellow of the Chartered Institute of Patent Agents.

A large number of the 800-odd patents agents are either partners or employees of firms of patent agents who deal with a wide variety of clients, ranging from a housewife with a new domestic gadget to great industrial organisations. However, as technology gets more and more specialised, individuals in the profession tend to specialise technically.

Where a company has a large amount of patent work it may be sound economics to have a full time patent department, with one or more qualified patent agents, in preference to using outside patent agents. In the latter case, one needs adequate arrangements for liaison with the agents. The point at which it becomes worthwhile to establish a patent department varies, dependent on the amount of patent work and its technical complexity. The trend with big companies nowadays is towards employing fully qualified patent agents full time.

20

The Banks Committee Report

Some years ago the President of the Board of Trade set up a committee, chaired by Mr. A. L. Banks, now Sir Maurice Banks, to investigate the British Patent System and to recommend alterations. Changes were and are needed because of the increasing rate of filing patent applications, which with greater range of technology delays the publication and grant of patents. The resulting uncertainties render planning difficult because of lack of knowledge as to whether there are patents in the offing which would create problems. Publication delays also cause duplication of research.

Another factor is the current trend to improve international co-operation on patents. We already have the International Convention for the Protection of Industrial Property, the main feature of which is that if one files a patent application in one member country and later files in another member country, the date of filing in the first country can be claimed as long as not more than one year has passed. This enables sale of products or discussion or exhibition of an invention to take place soon after the initial patent filing without filings in other countries being prevented. Another international co-operative effort exists in Scandinavia: Norway, Sweden and Denmark now have very similar patent systems and intend to make arrangements for one filing for what would be in effect a Scandinavian patent. Attempts are in progress for a comparable effort in the Common Market countries for a so-called European Patent. Finally, there is the Patent Co-operation Treaty, which aims to facilitate getting patents in more than one country on the basis of one filing and to reduce the amount of searching needed. Nowadays when an application is filed in several important countries they all do a search of the prior art; if one could rely on the search in one of a small batch of 'searching countries' this duplication of effort would be minimised.

In addition to these international matters, the requirements of the Strasbourg Convention were given emphasis in the Committee's deliberations. This was another agreement which sought to bring formal requirements for patent filings into more conformity than now exists.

These international trends had to be taken into account, since it would be undesirable to alter our system blindly and later to have to alter it again to meet international commitments. Whether we like it or not, we are no longer an island in the industrial sense.

In this country each application filed is examined for, *inter alia*, adequacy of description and novelty. However, the Patent Office cannot reject an application as not being inventive as compared with the prior art : all that is required is novelty. We all know that a new combination of integers can be obvious. If a granted patent claims something which, although novel at its priority date, is in fact obvious in the light of the existing state of the art at the priority date, the Courts can (on application from an interested party) declare the patent, or such of its claims as are obvious, to be invalid. The Courts form a second filter, albeit an expensive one. The thought behind this is that it is better to let too many weak applications mature into dubious patents than to risk the rejection of a good one by over-strictness at an early stage. This is in keeping with the basis of English criminal law, whereby it is held that the risk of several guilty people escaping is preferable to the risk that one innocent person might be wrongly convicted. In both cases the thought is laudable, but in patents it is now rather Utopian.

The major recommendations of the Report are early publication—eighteen months after the earliest priority date—and the separation of the search through the prior art and the examination of the application. This separation, it is said, would lead to many weak applications being abandoned when the search report is issued, thus reducing the Patent Office work load. It would also mean that the patent agent's responsibility in advising his client would be increased, as he would need to give guidance as to the extent of protection left after considering the search report, with the client having to decide whether further prosecution is warranted. Adoption of the other main proposals would call for a general increase in the standards of drafting of patent specifications.

One disappointing aspect of the Report is that Petty Patents or something like them are rejected. Such patents would, if permitted, be granted for ideas on the verge of obviousness, which at present lead at best to weak patents. They would be useful where the main reason for filing was to force publication, the chance of some protection being a useful 'free bonus'. They are available in several countries and usually have a shorter life than proper patents. They are in effect midway between patents for true inventions and designs registrations, where only features of appearance count.

It is hoped that if the recommendations are accepted and if they have the effects anticipated (two large ifs!) the Patent Office work load would be cut down and uncertainty reduced through speedier publication.

The most important points in the Report will now be discussed.

(1) Search and examination to be separated

At present the Patent Office Examiner does a search and issues an official letter including the search result if he finds prior art, and raising any other objections, e.g. lack of clarity of description, which may exist. Occasionally an application is accepted without the issue of an official action—this has happened to the author's efforts about six times in 22 years!

The Committee recommends that a search report be issued six months after filing the Complete Specification. Where the application was based on an overseas application under the International Convention or started with a Provisional Specification, this search report materialises 18 months after the earliest priority date.

Eighteen months after the earliest priority date, it is proposed, the Complete Specification should be published—unless in the meantime the application has been withdrawn. In a 'Convention' case or a Complete after Provisional case this would mean that the search report and the publication would occur at, or nearly at, the same time. The Report indicates that something more than merely laying the Complete Specifications open to inspection is visualised—probably printing.

On early publication, the applicant is to have the option of abandoning, or of seeking examination within one year of early publication if he still wants a patent. In the latter case a 'substantial fee' is visualized, to deter applicants from proceeding with weak cases. This may be good sense to a big company, which can afford such a fee for important applications, but it may be unfair to a small company or a private inventor. The latter may have a good invention needing money to develop, which he cannot get without a strong patent. If the fee is too great, he may not be able to continue with his application and thus lose the benefit of his invention. Thus, the idea of a large fee has a disadvantage. For a big company, if an application is filed for publication, splitting the search and the examination could help, as the application would then be abandoned on publication if the search report shows that little or no protection would result.

The period within which the application should be accepted is suggested as three and a half years from the earliest priority date, the Patent Office having discretion to refuse to proceed with an application where the applicant is excessively dilatory. This would avoid the practice of some people of dealing with an official letter on the last possible date!

These recommendations would speed up publication, which is good, and would mean, the Committee hopes, that many weak applications would not be examined. The Patent Office work load would thus be reduced, and perhaps that of patent agents. Early publication would give competitors notice of dangerous patents in the offing, which could reduce uncertainties in their business planning.

Early publication must come in some form—perhaps earlier than the Banks Committee visualises, and it is reasonable. Separation of search and examination, plus early publication, also look reasonable, apart from doubts about the substantial fee.

(2) The applicant is to be committed to the 'form of the specification as filed'

This would give much more limited scope for amendment than now possible, and would seem to prevent broadening of claims. At present under British practice, there is considerable scope for amending the description after filing and before acceptance. The principle is important, as its adoption would mean that claim and specification drafting would tend to improve, as would standards in revising cases originated overseas before filing. This applies where the specification, although satisfactory in its country of origin, is not satisfactory here due to differences in the legal systems.

(3) **Application files to be open to the public**

This follows USA practice, which is not necessarily good since conditions in the two countries differ greatly. Further, the propriety of the amendments made during the prosecution of the application would be open to question. This would call for care in dealing with official letters, as in what one said in letters to the Patents Office. Experience in the USA shows that unwise (or unlucky) comments in such correspondence can have damaging effects if referred to later.

(4) **Repeal of provisions for pre-grant opposition and extension of Patent Office jurisdiction on revocation**

At present the application is published on acceptance, and three months are allowed for an interested party to oppose the grant. The grounds on which these oppositions can be entered are less than the grounds on which one can apply to the Court for revocation of a patent. The Committee recommends that such oppositions be abolished, which looks a good idea as in general they are not very useful.

It is coupled with the recommendation that the Patent Office have jurisdiction to hear revocation proceedings throughout the life of a patent, and on the full range of grounds of revocation now available in the High Court, which are greater than those on which an application can be opposed. This would be cheaper than going to Court.

Of course, the Committee recommends that where an infringement action or an application for revocation is pending before the Court, an application for revocation to the Comptroller should not be allowed except with the leave of the Court.

(5) **After publication, a third party can initiate examinations or notify of further prior art**

It is recommended that an application to initiate examination be possible by a third party—e.g. someone who has good 'business' reasons for wishing to know what claims will be allowed. After such an application, proceedings would be between the applicant for the patent and the Patent Office, Hence, by paying a fee the third party would merely start the ball rolling. This could persuade the applicant to look at the third party's operations and attempt to slant his claims to give maximum annoyance to that third party. It may not, therefore, be wise to seek an examination on another person's applications.

Similarly, the Committee recommend that a third party can notify the Patent Office of further prior art, in which case the applicant is notified.

(6) Patent Appeal Tribunal to be replaced by a Patent Court)

This Court would handle patent (and one assumes designs and trade mark) matters, whether *de novo* or on appeal from the Patent Office. It would have at least three judges (we already have two patent judges who occasionally sit simultaneously in the same proceedings) and would have High Court status, right of appeal being to the Court of Appeal or occasionally, as provided for by Section 12 of the Administration of Justice Act 1969, direct to the House of Lords.

This seems reasonable and logical as it merely extends the established principle of having specialist judges and should reduce uncertainty and expense. The latter would follow if procedural measures also recommended are adopted. One important point here is the proposal that patent agents should be able to appear before the Patent Court; this would save the time and money now often spent on counsel.

Surprisingly, the use of technical assessors to advise the Court does not seem to have attracted much favour, as one would expect a positive recommendation to use such assessors where both sides agree. This could, if sensibly used, save much time and money otherwise spent on educating the judge. Recently in the Court of Appeal (in the appeal on the infringement action on the Valensi Patent) a technical expert was used to assist the Court.

(7) Statutory limits on the Patent Office search to be removed

At present the criterion of novelty is based on what is published in the United Kingdom with the proviso that certain documents, including patent specifications and official abridgements thereof which are over 50 years old, do not count as anticipatory matter.

The Committee recommends the removal of such limitations with the Comptroller having discretion on the extent of search.

Closely linked to this is the recommendation that an invention is new if it does not form part of the state of the art, i.e. everything made available to the public either by written or by oral description, or use, or any other way, before the priority date. This seems reasonable, as dissemination of information is quicker than it used to be, but it would give lawyers a new term—'state of the art'—to play with!

A further point which fits in with the above is the recommendation that the Patent Office be empowered to ask what prior art has been cited in other countries.

A final point is that where the applicant has a search done by the IIB (the International Searching Organisation at The Hague) the Committee recommends that this search be accepted as an official search, with remission of part at least of the applicant's fees.

(8) Comptroller to be able to reject an application on the ground of obviousness

As already mentioned, the Patent Office is at present unable to reject an application solely on the basis that one or more of its claims is obvious over the prior art. This is by contrast with many foreign countries where such powers exist.

The recommendation is 'trendy' and probably inevitable in view of the tendency to 'internationalisation'. It is reasonable if it is coupled with adequate training of examiners so that they can understand what is obvious to one skilled in the art, and if the Courts adopt a less rigid approach to claim interpretation. No reflection on the ability of examiners is intended; it is true, however, that most of them enter the Patent Office direct from university, so it is difficult for them to know how the man 'skilled in the art' really thinks. Furthermore, it is all too easy, as has been emphasised in the past by judges, to say, with hindsight, that an invention is obvious over the prior art. How does one know today what was, or should have been, obvious to one skilled in the art x years ago? In any case, obviousness is largely a matter of opinion.

On claim interpretation, at present the normal canons of interpretation of legal documents, which in this country are very strict, are applied. If a claim includes non-essential limitations, the claims may be interpreted as if such limitations were essential to the invention. In some countries where the Patent Office can reject an application as obvious, the Courts usually interpret claims more liberally than we do. In such countries a non-essential integer may be ignored by a Court considering infringement.

(9) The existing definition of invention to be retained, with specific inclusions and exclusions

The existing definition is, as mentioned above, fundamentally what was written over 350 years ago, plus case law on whether certain types of idea are legally 'inventions'. As a result of the Patents Act 1949, the definition was amended to include 'methods of testing applicable to the improvement or control of manufacture': the Committee were in favour of deleting this from the definition. If this is done, then methods of testing generally could well be mentioned in the inclusions list referred to below.

It is recommended that the old definition be retained, as it is difficult to formulate an adequate definition of an invention, and as any new definition would cause much litigation before it was fully interpreted, so that much of the case law would be scrapped. Retaining the definition as suggested would keep most of this case law, without introducing an area of uncertainty. Incidentally, in the Patents Act 1949, the word 'invention' is used in at least three different senses! Coupled with this it is recommended that a list of inclusions and exclusions from patentability is appended to the definition. Such lists, which should be readily alterable, e.g. by Statutory Instrument, would initially specify that agricultural and horticultural processes are patentable.

One interesting exclusion mentioned is that of computer programs. We can now patent these if we write the specification and claims in a suitable manner, although the patents thus obtained may be of dubious validity. The report regards programs as being more analogous to mathematical formulae than manufacture, showing the subconscious fixation on the object one can actually handle as a basis for a patent.

Some special form of protection for programs should be devised, difficult though this might be. However, in the absence of such special provision all we have is copyright (too narow in scope to be much use) and patents. Exclusion of protection of programs by patents therefore seems unfortunate. The Committee does, however, recommend that the Board of Trade keep the question of programs 'under review ... particularly in the light of international developments'.

(10) Shop right

It is recommended that where one person has used an invention before the priority date of the patent, but in a manner which would not upset that patent, he should be able to continue this use. This right would not be transferable, and would merely mean that the patent had a hole in its cover in respect of that person. Such shop rights, as they are often called, exist already in some countries. This could be useful if one could find a reliable way to prove prior use.

(11) Infringement

On this vital topic, the following are the main recommendations:

(a) *Contributory infringement*, which means selling parts for assembly to produce an infringing article, or selling machines or materials with the intention that they be used in an infringing process, would be actionable. This would in effect bring the decision in the old case of *Innes v. Short* (15 R.P.C. 449) into the law. In that case a patentee won an infringement action where a material was sold with instructions to use it in an infringing manner. The decision is regarded as bad law, although I feel it was morally right! The recommendation, if adopted, would make it easier to get adequate protection for some inventions.

(b) *A conspiracy to infringe* would be actionable. This is given a separate recommendation but in practice would often be a special case of (a).

(c) Specific provisions for actions for *declarations of non-infringement* would enable a manufacturer who wants to sell a device resembling what a competitor is selling to seek a declaration that it did not infringe his patents. He would have to apply initially to the patentee (or exclusive licensee) for such a declaration and get no response before bringing the action.

(d) *A definition of infringement* would be included in the Patents Act, and it is also to be stated that infringement can exist where the only differences between an article and claims are 'non-essential features of the claims'. If this means giving the claims a reasonably benevolent interpretation it could be good, but would make it more difficult to advise on infringement. It would be in effect a resuscitation of an old legal principle that you infringe if you take the pith and marrow of the other man's invention but differ from it in minor points.

(e) *Infringement damages* would be awardable from the early publication date if the acts complained of are within claims both as published and as granted, and the original claims are 'framed in good faith'. This, if adopted, should lead to better claim drafting.

(f) *Process or apparatus claims* would not cover the products: since one should always include 'product' claims whenever possible, with competent claim drafting this should make little difference.

(g) *Section 9 of the Patents Act 1949* would be repealed. This is the Section by which the Patent Office can insert a reference to another patent which would be infringed by use of one's own invention. The Section is erratic in its application, and has been largely made a nonsense by certain decisions of the Patents Appeal Tribunal. This is, therefore, an excellent suggestion.

(12) Conflict between concurrent applications

One major difficulty, especially where development is brisk, is that of concurrent patent applications, neither of which was published before the other, but which have or could have overlapping claims. Several recent decisions of the Courts and the Patent Appeal Tribunal have led to some confusion on this issue. The Committee recommends that the problem when it arises be settled on the 'whole content' basis, which would mean that an application of earlier priority date would be treated, as against a later-dated application, as if it was part of the prior art. This is seen as the least unsatisfactory solution to a difficult problem.

(13) Term of patent

This should be 20 years from the earliest priority date, as compared with the present 16 years from the filing of the Complete Specification. The renewal fees should be high for the last few years, which is reasonable. However, it is recommended that there be no extensions of patents for inadequate remuneration. This does not seem wise. Such extension should be available: there are not many cases where they have been granted, but they have been important ones. They include a patent on the Colpitts oscillator much used in radio, some early radio valve patents, the Valensi Patent on colour television and some of Sir Frank Whittle's jet engine patents: all of major importance. They should surely not be excluded from such a privilege.

The present 'war loss' provisions, by which extensions of the patent can be obtained if the patentee has been unable to exploit his invention due to a war in which this country is engaged, would be replaced by the granting of 'blanket' extensions by Order in Council. This may not be adequate, especially where the war has effects which precede and/or succeed it. In one application for an extension of term on this ground, Standard Telephones and Cables persuaded the Patent Office to take post-war circumstances into account in assessing the length of the extension. This was by no means the only case in which this was done.

(14) Revocation

Under the present Act, a claim can be held to be invalid if it covers anything lacking utility, even if it covers other things with utility. This would be replaced by a proviso by which the claim is only bad on this ground if it covers no useful embodiment. This would have the result that a claim covering something unworkable or useless, which can happen with the best patent agent, would merely be whittled down to useful things.

Other aspects of revocation have been commented on under the heading relating to the repeal of the existing provisions for pre-grant opposition.

(15) Disputes

Disputes between employer and employee, and possibly between an applicant and someone who feels that an application was made in fraud, would be settled by the Comptroller of the Patent Office, with, as usual, right of appeal to the Patent Court. The recommendations here are mainly of detail, and seem reasonable. However, if there are two applications and the applicants are in dispute, it is recommended that the Comptroller should be empowered where time is short to grant both, with a notice that a dispute exists. The dispute could then be settled, so to speak, at leisure.

(16) Compulsory licences

Repeal of the provision making compulsory licences mandatory for drug or food or medical-type patents is recommended. This provision is said to do little good in practice and may tend to inhibit drug firms from spending money on research. Adequate safeguards should exist in the provisions already in the Patents Act which relate to abuse of monopoly and to Crown rights to use a patented invention.

(17) Crown use

The provision for Crown use, whereby the Crown can use any patented invention, would be largely unchanged, but the payment for use would be on a willing licensee/willing licensor basis, which should help the patentee to get what he deserves. In assessing royalties the patentee's manufacturing interest would be considered, which would take account of the fact that a manufacturer would usually prefer to make and sell to the Crown than to license someone else, especially where that someone else is his rival.

(18) Employee invention

Very little change is recommended. However, it is suggested that the Department of Employment and Productivity should encourage voluntary schemes for rewarding inventors.

One disappointing feature of the Report is that no real consideration seems to have been given to how far the employer's interest extends when assessing whether an employee's invention belongs to the employer. This could cause trouble in large industrial organisations, where a group of companies has a very wide range of interests, from telecommunications equipment to, say, car brakes.

(19) Abridgements

The abridgements published by the Patent Office, which are now written by the Patent Office Examiners, are a valuable source of information, especially for searching into the prior art. They are much used for this by people outside the patent profession. The Committee suggests that this practice be replaced by the applicant supplying an abstract. This would be a retrograde step. In fact, the Patent Office suggested that this be adopted without delay, but the profession contested the suggestion—successfully for the time being.

The Banks Committee Report has been dealt with at length, but this is because it is a good Report, even if one cannot agree with all it recommends, and because its effects could be very significant on the British patent system. It also contains some interesting information about how the system works at present.

Bibliography

BLANCO WHITE, T. A., *Patents for Inventions*, Stevens, London (1962)

BLANCO WHITE, T. A., and JACOB, ROBIN, *Patents, Trade Marks, Copyright and Industrial Designs*, Sweet and Maxwell, London (1970)

BOARD OF TRADE, *Patents in the Common Market*, H.M.S.O., London (1970)

BOEHM, K., *The British Patent System. I. Administration*, Cambridge University Press, Cambridge (1967)

CHARTERED INSTITUTE OF PATENT AGENTS, *Patents Act* 1949-61, Sweet and Maxwell, London (1968). This is a commentary on the Act, and practice under it: supplements to it appear at regular intervals.

CHARTERED PATENT AGENT, 'Patents and the Engineer', *Electronics and Power*, September 1968, 358 *et seq.*

COPINGER, W. A., and JAMES, F. E. S., *Law of Copyright*, Sweet and Maxwell, London (1965)

FALCONER, D., ALDOUS, W., and YOUNG, D., *Terrell on the Law of Patents*, 12th edn., Sweet and Maxwell, London (1971)

KERLY, SIR DUNCAN MACKENZIE, *Law of Trade Marks and Trade Names*, Sweet and Maxwell, London (1966)

LEES, C., *Patent Protection, the Inventor and his Patent*, Business Publications, London (1965)

LIEBESNY, F., *Mainly on Patents*, Butterworths, London (1972)

MEINHARDT, P., *Inventions, Patents and Trade Marks*, Gower Press, Epping (1971)

NEWBY, F., *How to Find Out About Patents*, Pergamon Press, Oxford (1967)

PATENT OFFICE, *About Patents — Patents as a Source of Technical Information*, H.M.S.O., London (1971)

PATENT OFFICE, *Manual of Office Practice (patents)*, H.M.S.O., London (loose leaf)

PATENT OFFICE, *Searching British Patent Literature,* H.M.S.O., London (1970)

STEVENS, T. M., and BORRIE, G. J., *Mercantile Law,* 15th edn., Butterworths, London (1969)

STIGANT, S. AUSTEN, 'British Patents Procedure', *Electrical Review,* 23rd December 1960, 1078-1080.

WHALE, R. F., *Copyright,* Longmans, London (1972)

WHITE, W. W., and RAVENSCROFT, B. G., *Trademarks Throughout the World,* Trade Activities Inc., New York (loose leaf)

Index